LEARNING BY PROJECTS™

ARDUINO® REMOTE SENSING AND CONTROL

USING 433 MHz MODULES

Easy Data Transmission with the VWComm Library

LEARNING BY PROJECTS™

ARDUINO®
REMOTE SENSING AND CONTROL
USING 433 MHz MODULES

Easy Data Transmission with the VWComm Library

A Tutorial Collection
of Staged Projects

Remote Sensing and Control
with the Arduino® Microcontroller

STEVEN HODDER AND JOHN RUBIDGE

This edition published 2017 by:
Takahe Publishing Ltd.
Registered Office:
77 Earlsdon Street, Coventry CV5 6EL

Copyright © Steven Hodder & John Rubidge, 2017

ISBN 978-1-908837-06-6

The moral rights of the authors have been asserted

All rights reserved. This publication may not be reproduced, stored in a retrieval system or transmitted, in any form or by any means, electronic, mechanical, photocopying, recording, scanning or otherwise, without the prior permission of the publishers.

Disclaimer

While the general and technical information provided within this book has been subject to reasonable checks and verification, the authors and publisher give no undertaking in regard to the veracity, completeness, accuracy, or otherwise. Any electrical equipment including the Arduino and experiments suggested within this text are used with the express caution that any damage, injury, or loss involved is entirely attributable to the user. The authors and the publisher accept no responsibility for any such claims. Although websites and organisations are cited in this book, this is not an endorsement of these resources or the quality of information and/or goods available.

TAKAHE PUBLISHING LTD.

2017

To Christine and Julie

Also to our canine friends, past and present,

that have enriched our lives

Acknowledgements

The authors would like to thank Terry Wade for his detailed technical advice and helpful suggestions. Also, thanks to the Arduino community for the example sketches that they have provided in libraries and online, some of which form the basis for the sketches contained in this book. For the purposes of brevity, we have had to trim the comments that form part of the sketches listed within and so we apologise to those who have provided original content but are not explicitly acknowledged. The full sketches are available on the companion website and these are more fully commented with reference to the original authors where possible.

CONTENTS

Title	Page
Introduction	1
Writing Sketches	5
Experiment 1: Sending and Receiving Simple Data	15
Experiment 2: Sending and Receiving On/Off Signals	21
Experiment 3: Towards Reliable Signalling	29
Experiment 4: Communications Using Other Data Types	41
Experiment 5: Extending the Transmission Range	51
Experiment 6: Wireless Doorbell	63
Experiment 7: Simple Weather Station	75
Experiment 8: Remote Keypad Lock	93
Experiment 9: Using Multiple Receivers	103
Experiment 10: Remote Data Logging and Real-Time Clock	115
Final Remarks	127
Appendix A: Component List	129
Appendix B: The VWComm Library	133
Appendix C: Glossary of Terms	135
Appendix D: Standard Template Files	139
Bibliography	141

This book is part of the Learning by Projects™ series which use the same stage-by-stage approach to learning and development by concentrating on specific hardware options.

Introduction

This book introduces you to remote sensing and control projects using the Arduino Uno and the 433 MHz transmitter/receiver (Tx / Rx) modules. We have been quite specific in our coverage in an attempt to thoroughly explore the subject with particular reference to a pair of readily available and inexpensive remote control modules. These modules have a good transmission range and are simple to use – particularly with the library that we have created to simplify the sending and receiving of data.

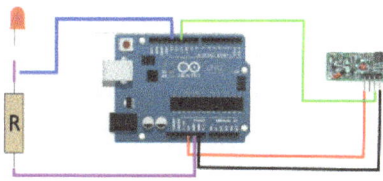

Assumed Level of Previous Knowledge

While this book is intended to be an informative source for a broad range of experimenters, it has been written with a narrow focus on a particular topic (remote sensing and control) using a particular resource (the 433 MHz modules). We expect that the reader will have conducted at least a few basic Arduino experiments and will have gained some familiarity with the programming language. We provide program notes for most experiments where we attempt to outline the purpose and working of the code. While we don't explain the program code in every detail, the programs are internally documented to indicate the purpose of relevant lines of code.

Why We Have Created this Book and How it Teaches You

There were two main thoughts behind this book: to investigate various remote control projects, and to do this with reference to a specific component through a series of related projects each involving a short set of experiments in which the user could gradually build up their knowledge and that additional programming code could be introduced in a readily understandable way. We wanted the reader to feel that having completed the book, they would be confident in setting up any project using these modules.

Basics of Remote Sensing & Control and Why it is so Useful

In recent years, we have seen the introduction of so many remote control applications. We were already familiar with the TV remote control, but now we accept as commonplace remote devices that unlock car doors, open garage doors etc. There are now many options to control such domestic settings as thermostats via a smartphone or to remotely view activity in your living room using the internet.

The Arduino is an ideal platform for remote sensing and control projects. With its many input/output capabilities it can receive data from a range of sensors and can send commands to devices to switch them on and off or send intermediate values. Also, there is a huge range of sensors and switches available at extremely low prices. These can usually be connected directly to the Arduino with a minimum of wiring and other components.

For remote sensing and control, we need to be able to transmit and receive data. There are several options available, but the cheapest and readily available are the 433 MHz Tx/Rx modules. These come as a pair for around £1 or as little as 50p when buying in quantity. They are relatively simple to use and require only three connectors (Vcc, Data, and Ground). Also the range of these devices is quite adequate for most home projects and they offer quite reasonable accuracy in transmission. For these reasons, we chose the 433 modules as the basis for this book so as to give the reader a comprehensive and practical guide to the modules and the uses to which they can be put.

Basic Details of the Standard 433 MHz Modules

The 433 Rx and Tx modules are shown above. The Rx has 4 pins which from the front, left to right are: Vcc, Data, Data, Gnd. It does not matter which data pin you connect to. Note that the pins are labelled on the back of the Rx module. The Tx module has 3 pins: Data (labelled ATAD), Vcc, and Gnd. Both can be connected to a 5V DC source.

To use these with the Arduino, you will need to include a library – this is essentially some code that had specially been written for a piece of hardware that simplifies its use. The most common library to use with these is Virtual Wire, although there is another one called RC Switch. Virtual Wire is a general purpose library that enables the user to send and receive a string of characters. RC switch is a more specialist library and is often associated with remotely controlling purpose-built devices. We wanted users to be able to write programs without needing advanced programming skills or technical knowledge. Accordingly, we decided to use Virtual Wire and write another library to sit alongside it providing some simple commands for sending and receiving data. This library is called VWComm and can be downloaded from this book's companion website (see below). More details of the Virtual Wire and VWComm libraries, how to install them and how to use them can be found in the relevant sections of this book.

As mentioned above, there is a companion website for this book:

http://www.learningbyprojects.uk

where libraries and sample code can be downloaded.

Further Details of the Standard 433 MHz Modules

The Rx operates with a supply of 5V, while the Tx can be powered with voltages between 3.3V and 12V. Higher voltages for the Tx can improve the transmission distance (see Experiment 5), but 5V is often convenient as this can be drawn from the same supply as that to the Arduino (although the Arduino can accept supply voltages between 5 and 12V, it might be advisable to not use maximum voltage ratings to the transmitter if the transmission times are likely to be lengthy and frequent).

These modules operate at a frequency of 433.92 MHz – midway in the so-called 433 MHz band which ranges from 433.05 MHz to 434.775 MHz and is divided into 69 channels, separated by 0.025 MHz. The 433 MHz band is also used by a wide range of other purposes including Amateur Radio, commercial remote control devices, and other users. This raises the possibility of interference but this generally isn't a problem and there are ways of minimising any such issues with the creation of software that recognises that such issues can occur, We shall look at these in the following chapters.

For convenience, all the experiments use the Arduino Uno for both Tx and Rx. These models have advantages for development purposes in that all input/output connections are provided by headers which enable standard connectors to be used. However, for practical purposes, the Uno might prove to be a bit oversized for some applications. We would therefore suggest the use of the Arduino Nano or the Micro which are more compact but still offer the same functionality. This probably is more applicable to the Tx which may be a remotely placed station for which size may be important.

Writing Sketches

We are assuming that you have a little programming knowledge and that you have installed the Arduino IDE. Check that your version is up to date at:

https://www.arduino.cc/en/Main/Software

Note that the Arduino web site also provides a good language reference guide at:

https://www.arduino.cc/en /Reference/HomePage

In the Arduino world, programs are referred to as *sketches* and are based upon the C programming language. Each sketch has a name ending in .ino and sits in its own folder of the same name. A good introduction is given in 'Programming Arduino – Getting Started with Sketches' by Simon Monk, ISBN 978-0-07-178422-1, published by McGraw-Hill.

This chapter is a quick reference guide to some of the main issues involved in typical Arduino programming. Remember that the example sketches that accompany the Arduino IDE can also provide a useful reference source.

Structure of a Sketch

All the sketches that you write will follow a similar pattern. Small variations are possible, but generally the code is set out in a particular order with comments included to make it more understandable. The value of comments should not be underestimated - they may seem a chore at the time, but will prove invaluable at a later date.

The structure of a typical Arduino sketch is shown below:

```
// Compiler Directives e.g.
#include <LiquidCrystal.h>

// Declarations e.g.
int count = 0;

void setup() {
// your setup code
}

void loop() {
// your main code
}
```

As shown above, a typical sketch has 3 sections:

```
directives/declarations
setup
loop
```

Directives are instructions to the compiler e.g. to include a library or to define a term as a constant e.g. #define ledPin 13. Unlike other code, these statements are not terminated with semicolons.

Declarations introduce variables that will be used in your sketch and, optionally, set them to some initial value. By *variables*, we refer to the storage of data under a name chosen by the user, the value of which can be changed during the course of the sketch. We shall look at various data types in Experiment 4 and how we can send values and text between remote devices.

The next section of the sketch is *setup*. This is an example of a *function* (see the later section for an explanation of *void* and the use of parentheses). The *setup* code will run once only and is the first sketch code to be executed. It is typically used to establish such things as inputs and outputs as well as starting and setting up certain associated utilities e.g. Serial.begin(9600) which initiates communication over the serial port at 9600 baud (bits per second).

The *setup* section is usually followed by *loop*. As the name implies, this is a section of code that will repeat indefinitely and is a typical requirement of micro-control programs. For example, we may want to continually read the state of an input from a sensor and activate a different response according to it being high or low.

Note that *loop* is optional, although *setup* will be needed to contain any code that will run once only.

The compiler ignores so called 'white space', so it does not matter if you include spaces where they are not strictly necessary (this can sometimes improve sketch readability). Text following // is treated as comment and ignored by the compiler. Sketches should be internally documented with comments to improve readability and aid debugging. A block of comments can be created by using /* to start and */ to end.

Braces, Parentheses, Semicolons, and Layout

Braces

Newcomers to the C programming language may be confused by the use of the curly brace { and }, particularly if they are used to programming in languages such as BASIC and its various forms. These symbols are only present to define a block of code that belongs to a sketch section or applies to some condition. At a top level, we have:

```
void loop() {
        // your main code
}
```

Or, at a lower level, we might have:

```
If (b == 0) { // check value of received data
        Serial.println("OFF");
}
```

The symbols define the beginning and ending of a block of code that is associated with some program structure.

We try to improve sketch readability by suitable indentation that emphasises the blocks of code (as well as adding comments to most lines of code). One very nice feature of the Arduino IDE is 'auto-format' which will adjust the layout of your code into a standard, readable format. It can be accessed under 'Tools->Auto Format' and is very useful in making your code easier to read.

Sketches can become complicated and it is easy to lose track of starting and closing braces. Note that placing the cursor at a curly brace will highlight its corresponding one – this can be very useful. The modern versions of the Arduino IDE automatically insert a closing brace when you type an opening one and this generally a helpful feature unless you are used to creating one yourself! A common compiler problem is a mismatch between opening and closing braces.

Semicolons

Every sketch command should finish with a semicolon e.g.

```
x = 5;
Serial.println(x);
```

These commands are not to be confused with *compiler directives* that appear at the start of a sketch e.g.

```
#include <VirtualWire.h>
#define mypasscode 27
```

These directives do not feature a semicolon at the end.

Layout

It is all too easy, particularly when writing code 'on the hoof' to neglect the proper layout of a sketch. If the compiler accepts it and the sketch works, why worry? But suppose that one month later, you find that your sketch malfunctions under certain conditions. You go back to your code and try and find the problem – the only trouble is that you've forgotten most of that logic that went into the design and all that you have is 100 lines of obscure sketch code. Where do you begin?

A little extra effort when writing a sketch will pay dividends at a later stage! The home enthusiast does not generally write extensive documentation for a sketch and, to be fair, we have all got better things to do. But we can make a compromise by laying out our sketch code in a way that clearly identifies the structure of the sketch and to include comments that explain what each section or line of code is doing.
To this end, we have already extolled the virtues of 'Auto Format' and have suggested the use of comments. It is helpful to include a commented section at the start of your sketch that explains any relevant details e.g. pin numbering.

Data Types and Declarations

Usually, we declare our variables in the initial section of our sketch (i.e. before the *setup* section). This creates so-called *global variables* that can be changed and accessed by any part of the sketch as opposed to *local variables* that might be defined within a routine but, even though they may have the same name, retain their individual status. It is common practice to declare variables within a sketch e.g. for(int I=0; i< 5; i++); although the authors would prefer that such variables be declared separately and at the start of the sketch. The declaration can also be used to set the initial value of a variable e.g. int i = 0; since it resets the value on every iteration of the loop! Beware that this can lead to unexpected consequences if used in the 'loop' section of a sketch.

Common Sketch Code

Many sketches can be written by using only a small subset of programming commands. Commonly used ones are outlined below:

Defining Inputs and Outputs

Examples: pinMode(readPin, INPUT);   pinMode(outPin, OUTPUT);

Note that readPin and outPin are names of constants defined at the start of the sketch e.g. const int readPin = 3;

Using Inputs and Outputs

Examples: v=digitalRead(readPin);   digitalWrite(outPin,HIGH);
Note that HIGH and LOW are pre-defined constants, equivalent to 1 and 0 respectively.

Conditional Statements

Examples: if(x==5){ .... } Note the use of the double equals sign when testing for equality. We can also use < (less than), <= (less than or equal to), > (greater than), >= (greater than or equal to), and != (not equal to).

Incrementing

Examples: i = i+1; (an alternative way to increment by 1 is i++;)

Iteration

Example: for (int i=0; i<5; i++){ .... } In this instance, the code within the curly braces is repeated 5 times, with i taking the values 0,1,2,3,4. Alternative options include while and do ... while.

Pauses

Example: delay(1000); causes the sketch to pause for 1 second (1000 milliseconds)

Other Programming Commands and Information

For a detailed list of structures, variables and functions see the Arduino Reference page as referenced at the start of this chapter. More explanation and examples can be found in Simon Monk's book also mentioned at the start.

Libraries

Many sketches make use of external libraries to simplify programming and provide convenient ways of using sensors, keypads, etc. A range of common libraries are included with the Arduino IDE and further ones can be added by downloading them from various sites including:

https://www.arduino.cc/en/Reference/Libraries

Libraries consist of several files, including examples, and can be downloaded as zip files. The Arduino IDE allows the user to easily add new libraries from the downloaded zip files from the menu: Sketch -> Include Library -> Add Zip Library ...

The authors of this book have created a library, VWComm, that greatly simplifies the sending and receiving of data and text with the 433 MHz Tx and Rx modules. This can be downloaded for free from the companion website for this book at: www.learningbyprojects.uk

Functions

A function is a section of code that lies outside the main body of code and can be invoked by its name, for example:

```
void flashLed() {
    for (int i=0; i<10; i++) {
        digitalWrite(ledPin, HIGH);
        delay(500);
        digitalWrite(ledPin, LOW);
        delay(500);
    }
}
```

This would be called from a line in the main sketch simply as: flashLed(); Functions can be useful if a particular section of code would otherwise be repeated several times within a sketch, or if they help to improve program structure and readability.

The word *void* indicates that the function does not return a value as opposed to say x = sqrt(y); where the function sqrt returns a value being the square root of the parameter y. Similarly, programmers can write their own functions which accept supplied parameters and return a single value of the declared function type. Some of the experiments in this book make use of user-defined functions.

The Serial Monitor

The Arduino is connected (sometimes temporarily) to a USB port on your computer to download sketches. This connection also allows the Arduino to send information back to the computer which can easily be displayed on the Serial Monitor on the Arduino IDE (click button on RHS of tool bar). We mentioned typical code at the start of this chapter.

When we download sketches, there is a small piece of software in the Arduino called a bootloader, that handles the installation over the USB connection. Note that the Serial Monitor also uses this connection and it should not be started until the sketch has been successfully downloaded. Running the Serial Monitor will automatically reboot the program. We should also be aware that digital pins 0 and 1 are used for serial communication and, since they share hardware connections with the USB and bootloader, they are best ignored for other purposes.

Development and Debugging

Many problems can be avoided by taking a systematic approach to your projects. Also, to avoid possible electronic conflicts from previous sketches, it would be wise to upload a sketch to the Arduino while it is disconnected from any circuitry, then remove the USB connection and make the necessary connections to your new circuitry. Always double-check the connections before running the sketch! Also, pay attention to your programming style and make sure that your code is readable, well laid out, and has a good amount of internal documentation. This may seem a pain at the time but when you come back to your code at a later stage, you will be glad that you took a little extra trouble in those early stages! Program readability can be greatly improved by the use of functions where appropriate. Not only can the code be more clear with a well-chosen function name, but the function can be tested separately to ensure that it does what was intended! Also, you can consider developing functions that can be used by other sketches that you might write in the future.

When you are creating a new project, it is tempting to write the complete sketches at one sitting. A more realistic approach, indeed one that is adopted throughout this book, is to develop the project in a series of steps in which you build up gradually in easy stages and testing as you go. Make sure you fully understand how things work and write test sketches with output to Serial Monitor.

However, having taken all these precautions, we often end up with a non-working project and this can be more tricky for remote control projects where two (or more) independent electronic bundles are involved. If nothing happens, you don't know whether the fault lies in the Tx or Rx (or both). The first step is to check the wiring and then to use the serial monitor to verify that the sketches are doing what you expected! Note that you can get unexpected results when both Tx and Rx are both connected to USB ports on the same computer unless they have distinct COM port numbers. For this book, we have found having additional (and proved) simple Tx and Rx stations standing by can be an invaluable weapon of last resource!

Using the Serial Monitor to trace what is happening in a program is a valuable resource and it can rapidly reveal programming errors. Common mistakes are misplaced end curly braces and the use of a single '=' in an if statement (as opposed to the double equal '==').

ARDUINO REMOTE SENSING AND CONTROL USING 433 MHz Tx / Rx MODULES

Experiment 1: Sending and Receiving Simple Data

Introduction

Many remote sensing and control projects involve sending only very limited information from the transmitter to the receiver. In this first experiment we shall establish that we can send simple data values from one Arduino to another using the 433 MHz Tx/Rx modules as described in the previous section along with the wiring details.

This experiment involves sending numbers to the receiver which displays them on a PC connected via a USB link (the display is known as the serial monitor). This forms the basis for sending more complicated data and how to respond to the received values.

Consideration of Technical Issues

You don't have to worry about the technical side of how values are sent and received because we are using the library 'Virtual Wire' and our own library routines for transmitting and receiving data values. The template sketches feature 'include' statements e.g. #include <VirtualWire.h> and #include <VWComm.h> that bring the relevant facilities into your sketch. For this experiment and others, you will base your sketches on the template Tx and Rx files that we have provided (see earlier section for downloads).

In this experiment, we send a value between 0 and 255 (a *byte* of data) and display it on the PC attached to the receiver via the USB link. This exercise might seem simplistic but it forms the basis of advanced monitoring and control projects and is an important first step in with 433 MHz devices.

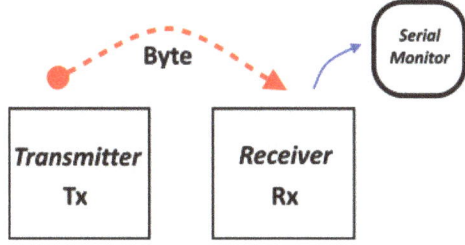

15

Construction Details

Connect your Arduinos as described in the earlier section of the book i.e. the 5V and 0V pins are connected to the pins Vcc and Ground (check that these are labelled as such!) and that the data pin is connected to 12 on the Tx and 11 on the Rx (note that these can be changed if you are also using another piece of hardware requiring these pins).

If you have two spare USB ports, you may have found that connecting a second Arduino changes the COM port in the first one! However, it is possible to have two Arduinos connected via separate USB ports and associated with different COM ports. Plug in the USB connections then launch the Arduino IDE sketches (***do not start the second IDE from one that is already open***). You will then be able to select the board and the COM port in each case.

Alternatively, you can upload the transmitter code first and power that part with a battery or power supply unit. The receiver should remain connected via a USB link so that it can send details to the serial monitor.

Sketch Details

The Tx sends a sequence of values in the range 0-255 (corresponding to 1 byte of data) to the receiver. The transmitter uses the function *sendByte(b)* and the receiver uses the function *readByte(b)*. For many remote control sketches, byte variables are sufficient for sending and receiving data. Given the limited memory size of the Arduino, they have the advantage of conserving storage space. We shall see, in the next experiment, how byte values can represent characters should we wish to send such things as "A" or "?". For now, we just wish to establish a basic communication between transmitter and receiver, so we will write a sketch that sends the values 0 to 255 with a short time delay between each value.

Both the Tx and Rx sketches will need to use the Virtual Wire and VWComm Libraries. Make sure that you have downloaded these (see Appendix 2 for details) and use the #include directive at the start of your sketches.

Sketch Code

The Transmitter Code is as follows:

```
/*
  Expt1_Tx - Transmitting sketch using Virtual Wire
  with a 433 MHz Tx module with VWComm library functions
*/

#include <VirtualWire.h>
#include <VWComm.h>
const int transmit_pin=12; // Tx data to D12
VWComm vwc; // create an object to use with VWComm
void setup()
{
  vw_set_tx_pin(transmit_pin);
  vw_setup(2000); // Data Tx rate (bits per sec)
}

void loop()
{
  for (byte b = 0; b < 256; b++) {
    vwc.sendByte(b); // send value in the range 0 - 255
    delay(1000);
  }
}
```

The Receiver Code is:

```
/*
  Expt1_Rx - Receiving sketch using Virtual Wire
  with a 433 MHz Rx module with VWComm library functions
*/

#include <VirtualWire.h>
#include <VWComm.h>
const int receive_pin=11; // Rx data to D11
byte b; // byte to hold received values
String dt; // string to use with VWComm
VWComm vwc; // create an object to use with VWComm

void setup()
{
  vw_set_rx_pin(receive_pin);
  vw_setup(2000); // Data Rx rate (bits per sec)
```

```
  vw_rx_start(); // start the receiver running
  Serial.begin(9600); // initialise serial output
  Serial.println("Starting...");
}
void loop()
{
  dt = vwc.dataType(); // wait for and identify data
  if (dt == "B") { // code 'B' indicates 'byte'
    b = vwc.readByte();
    Serial.print("Byte received: ");
    Serial.println(b);
  }
  else {
    Serial.println("Data type not recognised!");
  }
}
```

Testing the Sketches

First upload the Tx sketch and then remove the USB connection. Provide an independent power source to your Tx Arduino. Next, upload the Rx sketch to the other Arduino – you will probably need to change the 'Port' option in the 'Tools' menu of the Arduino IDE. Open up the Serial Monitor from the IDE and check that you are receiving the values. Test the range of your transmission by moving the Arduino Tx gradually away from the Rx – how far can you get without using any aerials? We shall discuss aerials in experiment 5.

Sketch Notes

The sketches in this experiment demonstrate the use of the VirtualWire and VWComm libraries to send and receive simple data values in the range 0-255. In further experiments we shall explore other common data types: text string, integer, and float. These provide enough facilities to provide easy communication between sender and receiver for a whole host of experiments.

It should be noted that both Tx and Rx sketches use some standard statements in the setup section to establish the relevant digital pins and to set the baud rate (bits per second). The Virtual Wire library defaults to the use of certain pins for various purposes, these are: vw_set_tx pin (defaults to 12), vw_set_rx_pin (defaults to 11) and vw_set_ptt_pin (defaults to 10). When setting up a new experiment, we should either avoid these pins for other purposes or change the default settings in the setup section. The ptt_pin (press to talk) is a transmitter enable pin whose default value is high when the transmitter is enabled. We don't use this in our experiments.

A common programming error is to use a single '=' instead of the double equal '==' when making a comparison, for example if (dt == "B"){ ......} The compiler will happily accept a single '=', but your sketch will give some strange results!

What You Have Achieved and the Next Steps

You have seen that you can successfully send simple data values between two Arduinos. While this may not seem a great thing, you have proved that remote communication works and you have paved the way towards creating complex control or monitoring projects.

In the next two experiments, we shall see how this basic idea can be extended to send and receive ON/OFF signals for switching purposes and to ensure that the Rx is ready to receive data before we start sending signals.

Experiment 2: Sending and Receiving On/Off Signals

Introduction

In the previous experiment, we learnt how to send bytes of data (values in the range 0 to 255) from a transmitter to a receiver using the Virtual Wire and VWComm libraries. This experiment involves four stages that extend the first experiment and demonstrates the use of 433 Tx / Rx modules for remote switching applications.

Consideration of Technical Issues

In this experiment, we start by sending a byte of data that will either be 0 or 1 to represent 'OFF' and 'ON' respectively. Initially, this will a continuous stream of 0s and 1s separated by a short time delay. The received value will be displayed on the PC attached to the receiver via the USB link. Once this has been achieved, we shall modify the receiver so that it displays "ON" or "OFF" and then add a push-button switch to the transmitter to toggle the on/off signals. Finally we add an LED to the receiver that is switched on and off from the remote transmitter switch.

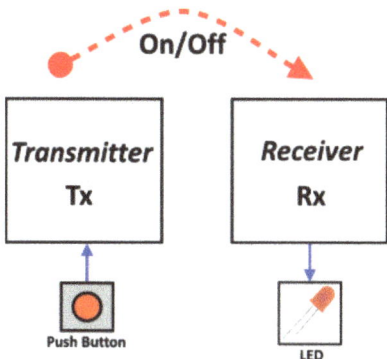

Construction Details (Stage 1&2)

Connect your Arduinos as in the previous experiment: 5V and 0V pins are connected to the pins Vcc and Ground, Tx data pin to 12 on the Transmitter and Rx data pin to 11 on the Receiver.

Stage 1 - Sending Simple ON/OFF Signals

Sketch Details

The Tx sketch requires only a small variation from the one used in the previous experiment. Instead of sending a sequence of values in the range 0-255, we shall only send 1s and 0s to represent on/off signals. For now, the Rx sketch can remain unchanged. This stage simply establishes the principle of sending and receiving on/off signals.

Sketch Code

Modify the Transmitter Code as follows:

```
/*
  Expt2_Stage1_Tx - Sending on/off signals as 1s or 0s
*/

#include <VirtualWire.h>
#include <VWComm.h>
const int transmit_pin = 12; // Tx data to D12
VWComm vwc; // create an object to use with VWComm
byte b = 1; // byte to hold transmitted values


void setup()
{
  vw_set_tx_pin(transmit_pin);
  vw_setup(2000); // Data Tx rate (bits per sec)
}

void loop()
{
  vwc.sendByte(b); // send an on/off signal as 1 or 0
  delay(1000);
  b = 1 - b; // invert the signal
}
```

Testing the Sketch

As before, first upload the Tx sketch and then remove the USB connection. Provide an independent power source to your Tx Arduino. Next, upload the Rx sketch to the other Arduino – you will probably need to change the 'Port' option in the 'Tools' menu of the Arduino

IDE. Open up the Serial Monitor from the IDE and check that you are receiving a sequence of 1 and 0 values.

Stage 2 - Receiving Simple ON/OFF Signals

We start with a small modification to the Rx sketch to interpret the 1s and 0s as on/off signals. This simply involves checking the received value and printing out ON or OFF to the serial monitor.

Sketch Code

Modify the Receiver Code as follows:

```
/*
  Expt2_Stage2_Rx - Receiving on/off signals as 1s or 0s
*/

#include <VirtualWire.h>
#include <VWComm.h>
const int receive_pin = 11; // Rx data to D11
byte b; // byte to hold received values
String dt; // string variable to use with VWComm
VWComm vwc; // create an object to use with VWComm

void setup(){
  vw_set_rx_pin(receive_pin);
  vw_setup(2000); // Data Rx rate (bits per sec)
  vw_rx_start(); // Start the receiver running
  Serial.begin(9600); // Initialise serial output
  Serial.println("Starting...");
}
void loop(){
  dt = vwc.dataType(); // wait for and identify data
  if (dt == "B") { // code 'B' indicates 'byte'
    b = vwc.readByte(); // get the byte of data
    Serial.print("Switch signal received: ");
    if (b == 0) { // check value of received data
      Serial.println("OFF");
    }
    else if (b == 1) {
      Serial.println("ON");
    }
    else {
      Serial.println("Data type not recognised!");
    }
  }
}
```

23

Testing the Sketch

Upload the Tx and Rx sketches as before and open up the Serial Monitor from the IDE of the Rx. Check that you are receiving a sequence of ON and OFF messages.

Stage 3 - Using a Push Button to Toggle ON/OFF Signals

After you have modified and tested the receiver code add the following circuitry to your transmitter set up: push-button on breadboard connected to digital pin 8 and GND, as shown:

The push-button will provide a digital input to pin 8. This pin will normally be HIGH until the button is pressed to temporarily make it LOW. To make sure that the digital input is normally HIGH, we could connect it to the 5V pin via a suitable resistor e.g. 10 KΩ. However, there is a simpler way that makes use of the Arduino's internal pull-up resistors. Having declared pin 8 as an input, we write a HIGH signal to the pin:

```
pinMode(inPin, INPUT); // set inPin as an input
digitalWrite(inPin, HIGH); // use internal pull-up
```

We are now going to add a push-button to the transmitter that will toggle the ON/OFF signals sent to the receiver. Pressing the push-button momentarily sends a LOW signal to the input pin which is constantly being scanned by the transmitting sketch (we refer to this as *'polling'*). The sketch initially sends a 0 to indicate OFF (within the setup section). When a button press is detected in the main loop, a 1 or a 0 is sent to indicate ON or OFF followed by a short delay. Each button push causes the signal to be inverted i.e. 1 -> 0 and 0 ->1.

Sketch Code

Modify the Transmitter Code as follows:

```
/*
  Expt2_Stage3_Tx  ON/OFF signals using button push
*/
#include <VirtualWire.h>
#include <VWComm.h>
const int transmit_pin = 12; // Tx data to pin D12
const int inPin = 8; // push button to pin D8
VWComm vwc; // create an object to use with VWComm
byte b = 1; // byte to hold transmitted values
void setup() {
  vw_set_tx_pin(transmit_pin);
  vw_setup(2000); // Bits per sec
  pinMode(inPin, INPUT); // set inPin as an input
  digitalWrite(inPin, HIGH); // unconnected = high
}
void loop() {
  if (digitalRead(inPin) == LOW) { // check button push
    vwc.sendByte(b); // send value of b
    delay(300); // short delay
    b = 1 - b; // invert the value of b (1->0, 0->1)
  }
}
```

Testing the Sketch

Upload the Tx and Rx sketches as before and open up the Serial Monitor from the IDE of the Rx. Check that you are receiving a sequence of ON and OFF messages. Note that we have included a short delay in the Tx after sending an ON/OFF signal to avoid multiple signals from a single keypress.

Stage 4 - Showing Received ON/OFF Signals with an LED

When you have successfully completed stage 3, add the following circuitry to your receiver set up: LED and current-limiting resistor on breadboard with wire to digital pin 13, as shown:

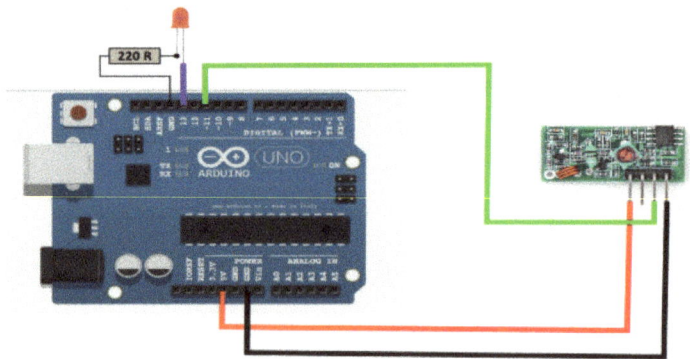

This final part of this experiment establishes that we can remotely switch a device. For this purpose, we do not anything complicated – if we can switch on an LED then we can switch on anything else! We modify the Receiver sketch to switch an LED on pin 13. The advantage of using this pin is that the Arduino has a built-in LED on that pin, so we can test or sketch before adding the external LED.

Sketch Code

Modify the Receiver Code as follows:

```
/*
  Expt2_Stage4_Rx   switching from ON/OFF signals
  Switch an LED ON or OFF
*/

#include <VirtualWire.h>
#include <VWComm.h>
const int receive_pin = 11; // Rx data to D11
const int ledPin = 13; // anode of LED to pin 13
VWComm vwc; // create an object to use with VWComm
String dt; // string to receive data type
byte b; // byte to hold received value

void setup() {
  vw_set_rx_pin(receive_pin);
  vw_setup(2000); // Bits per sec
  vw_rx_start(); // Start the receiver running
  pinMode(ledPin, OUTPUT); // set ledPin as an output
  Serial.begin(9600); // initialise serial output
  Serial.println("Starting..."); // print to serial mon
  digitalWrite(ledPin, LOW); // start in OFF state
}
```

```
void loop() {
  dt = vwc.dataType(); // wait for incoming data
  if (dt == "B") { // code 'B' indicates 'byte'
    b = vwc.readByte(); // get the byte of data
    Serial.print("Switch signal received: ");
    if (b == 0) { // check value of received data
      Serial.println("OFF");
      digitalWrite(ledPin, LOW); //switch LED off
    }
    else if (b == 1) {
      Serial.println("ON");
      digitalWrite(ledPin, HIGH); //switch LED off
    }
    else {
      Serial.println("Data type not recognised!");
    }
  }
}
```

Testing the Sketch

When you have uploaded the Tx and Rx sketches. Press the push-button on the Tx and verify that the LED on the receiver switches between ON and OFF at every press. With the Rx connected to a USB port, open up the serial monitor and check that the correct messages are being output.

Sketch Notes

Note that the final stage for the receiver sketch still contain commands to print information to the serial monitor. These are for verification purposes only and can be removed when the sketch is up and running correctly.

What You Have Achieved and the Next Steps

You can now use remote switching to turn a device on and off by sending simple signals. Although we are currently limited to switching an LED on and off, it is only a small step to extend this to other devices – a topic that we shall return to later in this book.

The current theme of sending and receiving data is developed in the next experiment, exploring possible problems and to attempt to find a way to establish reliable communication between devices.

Experiment 3: Towards Reliable Signalling

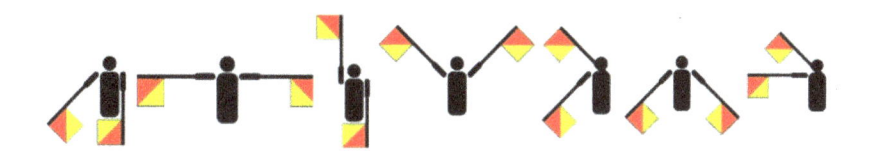

In the previous experiment, we learned how to send switching signals from a transmitter to a receiver using the Virtual Wire and VWComm libraries. These were simple ON/OFF signals where the TX sent an 'ON' signal as a value of 1 in a byte of data and similarly sent a 0 to indicate 'OFF'. We used these signals to toggle an LED connected to the Rx.

Generally this setup will work satisfactorily but it is vulnerable to various physical conditions that might result in unpredictable results. This experiment explores the problems that might occur and suggests simple ways around them. We also consider some other problems that require slightly more complex solutions and refer the reader to subsequent chapters that address these aspects.

Consideration of Technical Issues

Is the Rx ready to receive?

There would be no point in sending an ON/OFF signal if the receiver is not switched on or if it has not completed its initialisation. Similarly, the receiver cannot read another incoming signal until it has read the current data. At this stage, we shall introduce a simple visual check using an additional LED to indicate when the receiver is ready to receive a new signal.

Has the Tx responded to the push-button?

So far, the user cannot know whether a signal has been sent without looking at the receiver. Some visual indicator on the transmitter will confirm that the button-press has been found and acted upon. Also it will indicate when the transmitter is again ready to transmit. This can be achieved using two different colour LEDs or by using an RGB LED which would also enable us to indicate the data type being sent.

What are the issues concerning switch bounce?

We tend to think of push-buttons as simple, yet perfect devices that when you press the button, a connection is made which can make things happen. While this is basically true, there is a phenomenon known as 'switch bounce' that can cause some problems when responding to the switch press. The physical construction of the push-button means that instead of a single clean ON/OFF signal, we can get a rapid sequence of them (in a few milliseconds) before the final state is achieved. For some applications, this doesn't matter – particularly if the connections are to some physical devices e.g. a traditional doorbell. However, if the button is connected to a microcontroller (or certain other electronic components) which can detect and respond to the multiple on/off signals, these could result in multiple responses resulting in an unpredictable outcome.

It is usual to overcome switch bounce with the use of some simple electronic circuitry involving a capacitor and some resistors. A web search for 'switch debounce' will yield plenty of information on this method. This might be an important solution for remote control devices involving instant switching e.g. a model car, but for other situations involving interval switching there is a simpler solution. A short delay can be included once a switch push has been detected and this can prevent the detection of unwanted signals.

Situations where the user doesn't act as expected

Our previous test programs don't really demonstrate possible problems such as:

- Button is held down for too long
- Button is pressed multiple times when a single push is expected e.g. electronic doorbell (could use LCD display or use an LED to indicate ON or OFF + build-in code to ignore extra presses)

- Button isn't pressed hard enough

Discussion of the above issues

The push-button will typically be connected to the Arduino as shown:

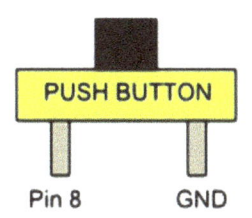

```
        PUSH BUTTON
        
        Pin 8    GND
```

Arbitrarily, we have chosen digital pin 8 for the input connection and, as described in previous experiment, we make use of the arduino's internal pull-up resistors to ensure that the unconnected state of the input pin is normally HIGH. Pressing the button will cause the pin to read LOW.

Generally, we detect the press of a button by constantly monitoring the state of the input pin using a technique called 'polling'. Typically, this involves using the 'loop()' section of our program where we read the input pin each time we go round the loop. If it goes low, we have a button press and we take some predetermined action. The following pseudocode illustrates the basic actions:

```
repeat:
      read input_pin
      if input_pin = LOW then
            send signal
```

However, as the processor can execute this code in a few microseconds, there is a likely problem because the user may not have released the button by the time the program instructions are repeated. This would cause several signals to be sent and this could produce unpredictable results at the receiver e.g. if the signal 'toggles' an output switch. Even if the transmitter sends alternate ON/OFF signals each time a LOW is detected, the problem still exists. The user may still be pressing the button for several iterations of the loop and this may be compounded by switch bounce.

These problems can usually be solved quite simply by thinking about both transmitter and receiver actions when designing the sketches. Techniques include using delays (if rapid switching is not required), sending definitive ON/OFF codes, using confirmation codes, or a combination of these techniques.

If our projects do not involve the remote control of toy cars, helicopters, etc. then we will generally be sending signals with relatively long time intervals and so we can design our software accordingly. For rapid transmitter actions, it may be necessary to incorporate switch de-bounce circuitry in addition to carefully designed software techniques. As the projects featured in this book are essentially non time-critical, we leave such aspects to the ingenuity of the reader (aided by the many articles available on the Internet).

User interface design for more complicated controllers

Almost all remote control projects will involve a user-interface at either or both ends of the transmitter and receiver. By this, we refer to such items as switches, LCD displays, LEDs etc. These are all ways in which the user interacts with the hardware and software that you have created. It should go without saying that this should be a simple and pleasant operation and yet it is often not so! Many commercial devices present obscure and difficult user interfaces that usually result in recourse to an equally obscure technical manual. This is very bad and totally unnecessary – the designers and manufacturers should consider some basic principles. For our purposes we should consider such aspects as:

Familiarity (similarity to commonly used interfaces)

- Simplicity (layout, adequate controls – don't overdo multiple-use functions)

- Confirmation (LED, LCD messages)

- Prompts (LCD messages)

- Aesthetics (colours, readability, layout)

It is beyond the scope of this text to describe these considerations in detail, but the above notes should be a useful hint. If the reader wishes to explore further aspects of user interface design, there are several useful websites such as:

http://bokardo.com/principles-of-user-interface-design/

although most of these are directed at designing graphical user interfaces for computer software.

Distance limitations

The 433 MHz Tx/Rx modules are usually supplied without aerials and, as such, the range is very limited. The signalling is unlikely to be reliable without aerials. In experiment 5, we shall look at the use of aerials and the options available. There are also 433 MHz modules with a longer range although these are more expensive than the ones used here.

The Experiments

Stage 1 – Adding an LED to the Tx for 'ready to transmit' and to the Rx to indicate 'ready to receive'

Sketch Details

These sketches have been derived from Experiment 2, and little modification is required. We shall still get the transmitter to send ON/OFF signals in response to a button push but we incorporate an LED that is normally ON, but goes OFF briefly while data is being sent. Similarly for the receiver, the LED is ON when waiting for data, but OFF while processing received data.

Sketch Code

The Transmitter Code as follows:

```
/*
  Expt3_Stage1_Tx Sending data values and
  adding LED indicator
*/
#include <VirtualWire.h>
#include <VWComm.h>
```

```cpp
const int transmit_pin = 12; // Tx data to D12
const int inPin = 8; // push button input to D8
const int ledPin = 9; // indicator for ready to transmit
byte b = 1; // byte to hold transmitted values
VWComm vwc; // Create an object to use with VWComm

void setup() {
  pinMode(ledPin, OUTPUT); // set pin for output
  digitalWrite(ledPin, LOW); // indicate 'not ready'
  vw_set_tx_pin(transmit_pin); // transmit pin
  vw_setup(2000); // Bits per sec
  pinMode(inPin, INPUT); // set inPin as an input
  digitalWrite(inPin, HIGH); //unconnected = high
  digitalWrite(ledPin, HIGH); // indicate 'ready'
}

void loop() {
  if (digitalRead(inPin) == LOW) {//check button-push
    digitalWrite(ledPin, LOW); // signal 'not ready'
    vwc.sendByte(b); // send value of b
    delay(300); // short delay
    b = 1 - b; // invert the value of b (1->0, 0->1)
    digitalWrite(ledPin, HIGH); // 'ready to transmit'
  }
}
```

The <u>Receiver Code</u> is as follows:

```cpp
/*
  Expt3_Stage1_Rx - Receiving on/off signals
  and indicating when ready to receive
*/
#include <VirtualWire.h>
#include <VWComm.h>
const int receive_pin = 11; // Rx data to D11
const int ledPin = 12; // connect anode of LED to D12
String dt; // string to received data type
byte b; // byte to hold received value
VWComm vwc; // Create an object to use with VWComm

void setup() {
  vw_set_rx_pin(receive_pin);
  vw_setup(2000); // Bits per sec
  vw_rx_start(); // Start the receiver running
  pinMode(ledPin, OUTPUT); // set ledPin as an output
  Serial.begin(9600); // initialise serial output
```

```
  Serial.println("Starting..."); // print to serial mon
  digitalWrite(ledPin, HIGH); // indicate 'ready'
}
void loop() {
  dt = vwc.dataType(); // wait for incoming data
  digitalWrite(ledPin, LOW); //indicate busy
  if (dt == "B") { // code 'B' - data type 'byte'
    b = vwc.readByte(); // get the byte of data
    Serial.print("Switch signal received: ");
    if (b == 0) { // check value of received data
      Serial.println("OFF");
    }
    else if (b == 1) {
      Serial.println("ON");
    }
    else {
      Serial.println("Data type not recognised!");
    }
  }
  digitalWrite(ledPin, HIGH); // indicate 'ready'
}
```

Stage 2 – Adding an LED to the Tx to indicate 'data sent'

Sketch Details

This is a simple modification to the Tx code in Stage 1 and does not involve any change in the wiring. As an alternative to indicating 'ready to transmit', we use the LED to indicate 'data sent'. The LED is normally OFF, but once data has been sent, it is held ON for two seconds.

Sketch Code

Within setup(), change the last line from:

```
digitalWrite(ledPin, HIGH); // indicate 'ready'
```

to:

```
digitalWrite(ledPin, LOW); // indicate 'waiting'
```

and change the code within loop() as follows:

```
void loop() {
  if (digitalRead(inPin) == LOW) { // check for btn push
    vwc.sendByte(b); // send value of b
    delay(300); // short delay
    b = 1 - b; // invert the value of b (1->0, 0->1)
    digitalWrite(ledPin, HIGH); // 'ready to transmit'
    delay(2000);
    digitalWrite(ledPin, LOW); // indicate 'waiting'
  }
}
```

Stage 3 – Using an RGB LED to indicate 'data sent'

Introduction

In stage 2, we added an LED to give a simple indication that data had been sent. As a refinement, we could replace the standard LED with an RGB LED and then indicate what sort of data has been sent, or its value. This is perhaps an unnecessary refinement, but it introduces the use of a function to simplify and shorten program code. At this stage, we are not exploiting the full capabilities of the function because we are only selecting a blue light to indicate when we have sent a byte of data. However, in Stage 4, we shall extend the use of a simple 'if' statement to include multiple 'else if' parts in the Tx code and make the LED illuminate from a range of pre-determined colour options.

Sketch Details

The program uses a function to set the LED colour. This reduces the program code because, otherwise, we would have 3 lines instead of 1 and the code would not be as readable. The term 'void' indicates that the function does not return a value.

Sketch Code

```
/*
  Expt3_Stage3_Tx use RGB LED to indicate
  byte of data sent
*/
#include <VirtualWire.h>
```

```
#include <VWComm.h>
const int transmit_pin = 12; // Tx data to D12
const int inPin = 8; // input from push button
const int redPin = 7; // LED 'red' pin
const int greenPin = 6; // LED 'green' pin
const int bluePin = 5; // LED 'blue' pin
byte i = 0; // byte to hold value to be sent
VWComm vwc; // create an object to use with VWComm

void setup() {
  vw_set_tx_pin(transmit_pin);
  vw_setup(2000); // Bits per sec
  pinMode(inPin, INPUT);
  digitalWrite(inPin, HIGH); // use pull-up
  pinMode(redPin, OUTPUT);
  pinMode(greenPin, OUTPUT);
  pinMode(bluePin, OUTPUT);
  setColour("OFF"); // switch LED off
}

void loop() {
  if (digitalRead(inPin) == LOW) {
    i = 1 - i; // toggle data value
    vwc.sendByte(i); // send data
    setColour("BLUE"); // switch on LED (blue)
    delay(2000); // delay 2 seconds
    setColour("OFF"); // switch off LED
  }
}

void setColour(String ledCol) {
  // set all led pins low
  digitalWrite(redPin, LOW);
  digitalWrite(greenPin, LOW);
  digitalWrite(bluePin, LOW);
  // then set pins to the chosen colour
  if (ledCol == "BLUE") {
    digitalWrite(bluePin, HIGH);
  }
  else if (ledCol == "GREEN") {
    digitalWrite(greenPin, HIGH);
  }
  else if (ledCol == "RED") {
    digitalWrite(redPin, HIGH);
  }
  // add other colour options here
}
```

Stage 4 – Indicating the Value of Data Sent

Sketch Details

The sketch will send a different data value on each press of the push-button. We choose the following colours for the various data values:

1:	●	blue
2:	●	red
3:	●	green
4:	○	white

To display 'white', we need to set each led output pin to HIGH. This will be done in the function setColour.

Sketch Code

In function setColour, replace the comment:

```
// add other colour options here
```

with the following code:

```
else if (ledCol == "WHITE") {
   digitalWrite(redPin, HIGH);
   digitalWrite(greenPin, HIGH);
   digitalWrite(bluePin, HIGH);
}
```

The loop code is modified as follows:

```
void loop() {
  if (digitalRead(inPin) == LOW) {
    i = i + 1;
    if (i > 3) {
      i = 0;
    }
    switch (i) {
      case 0:
        vwc.sendByte(i);
        setColour("BLUE");
```

38

```
        break;
      case 1:
        vwc.sendByte(i);
        setColour("RED");
        break;
      case 2:
        vwc.sendByte(i);
        setColour("GREEN");
        break;
      case 3:
        vwc.sendByte(i);
        setColour("WHITE");
        break;
    }
    delay(2000);
    setColour("OFF");
  }
}
```

What you have Achieved and the Next Steps

In this experiment, we have considered some of the problems that might occur in a remote control project and presented possible ways of avoiding them. We have seen that it is important to ensure that the transmitter is ready to send before we attempt to use a control action. Similarly, the receiver must be ready to read an incoming signal. If the project involves human action at either transmitter or receiver, then we must ensure that we communicate with the user so that they know what to do, when to do it, and to confirm that the required action has been taken. The last two stages of this experiment introduce some program concepts: a function and the switch/case selection.

The discussion here is by no means exhaustive, but it aims to give the reader some indication of things to consider when devising a project. A possibility for a further experiment might be to use a Tx/Rx pair of modules at both ends so that the stations can communicate with each other so that, for instance, confirmation of actions can be achieved. However, there will be a need to ensure that both Tx stations are not transmitting at the same time.

ARDUINO REMOTE SENSING AND CONTROL USING 433 MHz Tx / Rx MODULES

Experiment 4: Communications Using Other Data Types

Introduction

In the previous experiments, we sent simple signals using the Virtual Wire and VWComm libraries to switch a device ON or OFF and considered ways of ensuring reliable communication. The simple signals consisted of 1s and 0s and these are sufficient for many basic control applications. However, there are many other situations when we wish to send more advanced data values. For example, a remote weather station might want to send temperature and humidity values, alternatively it might interpret sensor data and send information in text form. Fortunately, the VWComm library makes this process very simple.

In this experiment, we shall consider the available data types and the various stages will introduce each one in turn. We look at any limitations on the data that can be sent and offer alternatives for the rare situations that breach these conditions. Programming issues are also considered where different data types are involved.

Consideration of Technical Issues

Why are there different data types?

The values that we may wish to send fall into three categories: byte, integer, and float. We shall also look at sending short text messages. In our everyday usage, we make no distinction between the type of data values but computers store these values in different ways. We also have to be aware of this when we are doing calculations or making comparisons, otherwise we may get some unexpected results!

The different data types have different storage requirements and, as the Arduino, has limited storage capabilities it is often important to choose a data type that uses the least storage for its purpose.

We shall also mention the use of arrays, which enable you to store a collection of similar data types with a single name. These can be thought of as a list where each item is identified by a number representing its position in the list.

The Arduino programming language is C++ which also enables the user to store text. The Virtual Wire library allows you to send and receive text, but the underlying code can be a bit daunting for those who are not experienced C++ programmers. We make use of another option - the String data type - which is simple to use and we let the VWComm library handle the more difficult bits! The length of the text is quite limited, but this should not be a problem with sending and receiving the sort of messages that we typically use for remote sensing and control purposes.

The Experiments

Stage 1 - A Generalised Receiver Sketch

For the various stages in this experiment we shall be writing transmitter programs for the various data types. To simplify things, we shall start off by creating a receiving program that will respond to all the data types in the VWComm library and output results to the serial monitor. At this stage, we ask the reader to accept the Rx sketch without full explanation - relevant details will be given in the following stages where the individual data types are described in detail.

Sketch Notes

This sketch is only an extension of the one used in previous experiments and makes use of the VWComm library. The information received from the transmitter is in two parts: the data type and the actual data. It is important to understand how the receiver sketch identifies the incoming data type and then how it deals with the incoming data value.

A call to the function dataType as in 'dt = dataType()' causes the program to wait for incoming data. The first part is a single byte that contains the ASCII code for 'B', 'I', 'F' or 'S'. This tells the sketch how to deal with the second part - the actual data value.

The VWComm library makes it easy to identify and receive different data types. This makes it simpler for the programmer to write code for a variety of remote control/sensing situations. The following code illustrates how to detect and respond to different data types and this will be used throughout the various stages of this experiment. You can test this using the Tx code from any stage in Experiment 3.

Sketch Code

The Receiver Code is as follows:

```
/*
   Expt4_Stage1_Rx - Generalised Receiver Sketch
*/
#include <VWComm.h>
#include <VirtualWire.h>
const int receive_pin = 11; // Rx data to D11
String dt; // string to received data type
byte b; // byte to hold received value
int i; // integer to hold received value
float f; // float to hold received value
String text; // string to hold received text
VWComm vwc; // create an object to use with VWComm

void setup() {
  vw_set_rx_pin(receive_pin);
  vw_setup(2000); // Bits per sec
  vw_rx_start(); // Start the receiver
  Serial.begin(9600); // initialise serial output
  Serial.println("Starting...");

}

void loop() {
  dt = vwc.dataType();
  if (dt == "B") {
    b = vwc.readByte();
    Serial.print("Byte received: ");
    Serial.println(b);
  }
  else if (dt == "I") {
    i = vwc.readInt();
    Serial.print("Integer received: ");
    Serial.println(i);
  }
  else if (dt == "S") {
```

```
      text = vwc.readStr();
      Serial.print("String received: ");
      Serial.println(text);
   }
   else if (dt == "F") {
      f = vwc.readFloat();
      Serial.print("Float received: ");
      Serial.println(f, DEC);
   }
   else {
      Serial.println("Data type not recognised!");
   }
}
```

Stage 2 - Whole Number Values: Byte and Integer

Sketch Notes

For many control applications, the values that we send and receive are whole numbers. For example, to select just one of several devices, we would probably send a number to identify that device. We have already seen how we send 1s and 0s to indicate 'ON' and 'OFF' values. In other situations, whole numbers may be quite adequate e.g. a temperature in degrees Celsius.

Essentially, when we are dealing with whole numbers, we can choose between two data types: byte or integer. A byte can store values between 0 and 255 whereas an integer can store values between -32767 and + 32767. As the name implies, byte occupies just a byte of storage whereas an integer uses 4 bytes. Note that there is also a long int type that supports an even greater range of values using 8 bytes, but this is not supported in the VWComm library.

Care must be taken to use the correct data type when sending data. Sometimes your data may be of a different type - for example, you may be receiving decimal values from a temperature sensor and yet you are only interested in using whole numbers. There are program functions that will perform the necessary conversions for you e.g. i = int(t) or b = byte(i). Note that you cannot squash a value bigger than 255 into a byte without losing something! For example, if the integer variable i contains the value 400, then byte(i) will return a value of 144 (the 8 least significant bits of i).

When writing sketches for the Arduino, we often find it convenient to use the built-in data values of ON/OFF and true/false (note these are lower case). These data values are actually boolean variables that hold 1 and 0 respectively and can be used in calculations if necessary. If you want to send these values using VWComm, treat them as byte data type.

Sketch Code

```
/*
    Expt4_Stage2_Part1_Tx  -  sending bytes as 1s and 0s
*/
#include <VirtualWire.h>
#include <VWComm.h>
const int transmit_pin=12; // Tx data to D12
byte b=1; // byte to hold transmitted value
VWComm vwc; // create an object to use with VWComm

void setup() {
vw_set_tx_pin(transmit_pin);
vw_setup(2000); // Bits per sec
}

void loop() {
    vwc.sendByte(b); // send value of b
    delay(2000); // short delay
    b=1-b; // invert the value of b (1->0, 0->1)
}
```

Note that if, for example, we want to send boolean values we would make the following changes to the sketch:

In the declarations: `boolean b=true;`

In the loop: `b=!B; // invert the value of b (T->F, F->T)`

Now try sending integer values. In the declarations, replace that for b with: int i = 250; and replace the code in the loop with:

```
        vwc.sendInt(i); // send value of i
        delay(2000); // short delay
        i=i+1; // increment i
```

Stage 3 - Decimal Values: Floating Point

Sketch Notes

Floating point numbers are essentially those with a decimal point e.g. 5.75 as opposed to whole numbers. They can be positive or negative between approximately $\pm 3.4 \times 10^{-38}$ and $\pm 3.4 \times 10^{38}$ (very tiny numbers to very large numbers, either positive or negative). In programs, we declare them as float e.g. float x; or float y = 2.75;

The Arduino stores float variables in 4 bytes (32 bits) although not all of this holds the actual digits of the number. We can only expect an accuracy of 6 significant digits i.e. the digits starting from after the leading zero. For example, 438764297 might get stored as 438764000 and 0.00327168156 might get stored as 0.00327168

Some care needs to be taken with programs using float variables because parts of calculations might get treated as integer values unless a decimal point is specified. For example:

```
float x=3;
x=x/2;
```

gives the correct result (1.5) whereas:

```
float x=3;
x=x/2 + 1/3;
```

gives the same result (1.5) because 1/3 is processed as an integer calculation resulting in 0.

However, the correct result can be achieved by including a decimal point e.g.

```
x=x/2 + 1./3;
```

To be on the safe side, specify all values with a decimal point unless you really want to use integer values e.g.

```
x=x/2.0 + 1.0/3.0;
```

Care should be taken when using floating point values as loop control variables, for example for(f=0.0;f=2.0;f=f+0.1) will not stop as expected when f equals 2.0 because of the slight error in the way that the increment of 0.1 is stored in a finite number of bits.

Sketch Code

```
float f = PI; // float to hold transmitted value
    :
    :
void loop() {
    vwc.sendFloat(f,4); // send value of f
    delay(2000); // short delay
    f=f+1.8; // increment f
}
```

Note that the sendFloat command has two parameters: the variable and the number of decimal places. At this stage, the VWComm library does not receive the number of decimal places and the receiving program should be written with this in mind. This is only relevant for displaying the received values, typically using the serial monitor or an LCD display. Serial.print(x) defaults to 2 decimal places but you can use Serial.print(x,N), similarly with LCD.print().

Stage 4 - Text Values: String Type

Sketch Notes

Although most practical applications involve sending values to the receiving station, there are some situations when we wish to send text messages. We can store text in two ways: as a character array, or as a 'String' type. It is easier to use the String type and there are plenty of associated functions that enable you to do various things with them e.g. find its length, get a substring, remove leading or trailing spaces, and so on. See the Learning Reference section of the Arduino home site:

https://www.arduino.cc/en/Reference/StringObject

VWComm uses Virtual Wire for sending the data and Virtual Wire places a limit on the maximum number of characters to be sent in any one go. VWComm will automatically truncate any strings, but we would rather avoid this as we want to be sure of what we are sending. Although strings will be limited to 27 characters, this will probably be sufficient if we want to display the received text on an LCD display and we could, if necessary, send more than one string.

Sketch Code

The sample code for this stage (part 1) can be as simple as shown below (following the usual declarations and setup).

```
void loop() {
    vwc.sendStr("Hello World!"); // send message
    delay(2000); // short delay
}
```

Alternatively, we could declare a string variable at the start of the sketch:

```
String myText = "Hello World!";
```

and then change the line in the loop to read:

```
vwc.sendStr(myText); // send message
```

Note that you should use the double inverted commas for the String type - you will see the colour of the displayed text is blue in the Arduino IDE when the correct usage is applied. Note that the IDE's use of colour can be very helpful when debugging!

Putting it all together

As a final part to this experiment, let's bring together the various data types and write a sketch that involves sending a different type at the press of each button. The generalised receiver sketch used throughout these experiments can still be used to display incoming data on the serial monitor.

```
/*
  Expt4_Stage4_Part1_Tx Sending different data types
*/

#include <VirtualWire.h>
#include <VWComm.h>
const int transmit_pin = 12; // Tx data to D12
const int inPin = 8; // input from push button (D8)
byte i = 0; // byte to hold transmitted value
float x; // float to hold transmitted value
VWComm vwc; // create an object to use with VWComm

void setup() {
  vw_set_tx_pin(transmit_pin);
  vw_setup(2000); // Bits per sec
  pinMode(inPin, INPUT);
  digitalWrite(inPin, HIGH); // use pull-up
}

void loop() {
  if (digitalRead(inPin) == LOW) {
    i = i + 1;
    if (i > 3) {
      i = 0;
    }
    switch (i) {
      case 0:
        vwc.sendByte(i);
        break;
      case 1:
        vwc.sendInt(i + 256);
        break;
      case 2:
        x = float(i) * 15.0 / 7.0;
        vwc.sendFloat(x, 4);
        break;
      case 3:
        vwc.sendStr("This is text");
        break;
    }
    delay(2000);
  }
}
```

What you have Achieved and the Next Steps

In this experiment you have studied the difference between various data types and learnt how to program the 433 modules to send and receive them.

You have used a transmitter program to send all the different data types: byte, integer, float, and string to a generalised receiver sketch.

You might wish to extend the transmitter sketch to include the various coloured LED signals that we described in Experiment 3 to indicate the type of data being sent - we leave this as an exercise.

Typically, you will use these to transmit data from various remote sensor modules and it will be necessary to find out what type of data your sensor is providing so that you can use the appropriate program code. It is generally possible to convert from one data type to another, although some loss of detail may be involved e.g. in converting a floating point number to an integer. You will need to modify the generic Rx sketch according to your particular requirements and this may involve an LCD display, relay switching, LED indicators etc.

In the following chapters, we take you through a series of experiments that demonstrate some of the practical capabilities of the 433 Tx/Rx modules when combined with additional electronic components and using the VWComm library to send and receive data.

Experiment 5: Extending the Transmission Range

Introduction

Without connecting any sort of aerial to the Tx/Rx modules, the transmission distance is typically less than 2 metres which is unrealistic for most practical applications. The range can be extended to 10 or more metres with a suitable aerial on both the transmitter and receiver. Ideally, transmission will be 'line of sight', but signals can pass through walls although the overall range will be reduced.

In this experiment we suggest some basic tests and then look at various aerial options, including home-made and commercial alternatives. We also consider the likely ranges attainable as well as other factors that affect the transmission range.

Consideration of Technical Issues

The Tx/Rx modules operate at a frequency of 433 MHz which is at the lower end of the UHF band (300 MHz to 3 GHz). This band is also used by numerous other devices e.g. television, Bluetooth, cordless phones, GPS etc. Transmitters and receivers need to use aerials (or antennas) for reliable communication and these can be rather large for low frequencies. However, for the UHF range, the aerials are quite compact and this is an important factor for portable devices. The main disadvantage of using higher frequencies is that the transmission range is shorter and limited to 'line of sight' communication (at lower frequencies, signals can be bounced off the ionosphere and achieve considerable distances).

There are three ways to increase the range of the 433MHz modules:

1. *Decrease the baud rate (the number of bits per second)*
2. *Raise the voltage of the transmitter and receiver. Most transmitters have an operating voltage of between 3V and 12V and the receiver 5V.*
3. *Attaching a suitable aerial – practical options are examined in this chapter.*

Aerial theory is complex and also involves a lot of trial and error. In general, for the 433Mhz modules a ¼ wavelength aerial is the most practical although a ½ wavelength is likely to give better results, but is less convenient. Using the formula Wavelength = 300/Frequency (MHz) the length of a ¼ wavelength aerial should be 173 mm. Alternatively, a ½ wavelength aerial would have a length of 346 mm.

The Virtual Wire library allows the user to set baud rates for the transmitter and receiver (which must be the same). A useful source of information about Virtual Wire can be found at:

https://www.seeedstudio.com/depot/images/product/VirtualWire.pdf

This suggests that realistic baud rates can be set between 1000 and 7000 bps. The range increases with a decrease in baud rate, although there is little to be gained below 2000 bps. Baud rates above 7000 bps may result in transmission errors and the range can be nearly halved. The projects described in this book don't require fast baud rates and we suggest that 2000 bps is an appropriate setting.

The Experiments

The general theme of these experiments is to obtain some idea of practical transmission distances through a series of small changes involving Tx voltage, baud rate, aerials, and obstacles. For this purpose, we want to observe the quality of the received signal, while changing the distance between Tx and Rx or introducing obstacles between them. The easiest way of doing this is to fix the position of the Rx and connect it to a computer and view the results on the serial monitor.

The Tx should be made as portable as possible, so a battery powered Arduino would be a good solution. If you have a 'Nano', then it and the Tx module can be mounted on the same board with a PP3 battery via a power regulator module.

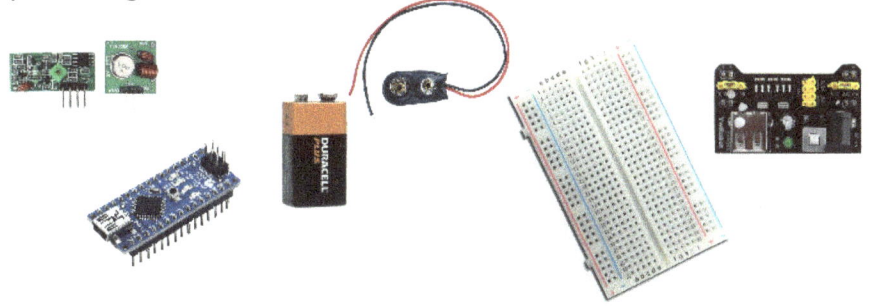

Stage 1 – Basic Tests without an Aerial

This stage aims to establish the basic range of the 433 modules and to discover what improvements can be achieved by first increasing the operating voltage of the transmitter and then varying the baud rate. A simple test of transmission distance can be made by sending a sequence of signals from the transmitter to the receiver which display the received values on the serial monitor. By gradually moving the transmitter or receiver further away, we can measure the maximum distance. We shall also experiment with the baud rate and the supply voltage to the transmitter. For consistency, the Tx and Rx modules should maintain their orientation throughout e.g. front of modules facing each other. Also, the experiments should be conducted on a flat, clear and level surface. Even so, we cannot eliminate small variations and results are perhaps best quoted from an average of measured values taken at different times. If the individual measurements differ significantly, the conclusions may be questionable!

Sketch Notes

In experiment 4, we explored the sending and receiving of different data types, so a suitable test sketch could be based upon this – see the example sketches supplied with VWComm. The variety of data should exploit any variation in transmission reliability and provide a reasonably accurate estimation of the reliable transmission distance by observing the received data values. To experiment with different baud rates, change the line for Tx and Rx in setup that reads:

```
vw_setup(2000); // baud rate (bps)
```

Finally, we can try increasing the Tx voltage supply. We start off by measuring the transmission distance when the Tx is connected to 3.3V and then repeat the exercise with 5V offered by the Arduino. We then use an external power source to examine the difference obtained with 9V and the limit of 12V supply to the Tx module.

Sketch Code

The Transmitter Code is as follows:

```
/*
  Expt5_Tx - Transmitting Sketch
  Send different data types repeatedly
*/

#include <VirtualWire.h>
#include <VWComm.h>
const int transmit_pin = 12; // Tx data to D12
VWComm vwc; // create an object to use with VWComm

void setup()
{
  vw_set_tx_pin(transmit_pin);
  vw_setup(2000); // baud rate (bps)
}
void loop()
{
  vwc.sendByte(7);
  delay(1000);
  vwc.sendStr("This is a test");
  delay(1000);
  vwc.sendInt(1234);
  delay(1000);
  vwc.sendFloat(752.1415926, 4);
  delay(1000);
  vwc.sendStr("....");
  delay(1000);
}
```

The <u>Receiver Code</u> is as follows:

```
/*
   Expt5_Rx - Generalised Receiver Sketch
*/

#include <VWComm.h>
#include <VirtualWire.h>
const int receive_pin = 11; // Rx data to D11
String dt; // string to received data type
int i; // integer to hold received value
byte b; // byte to hold received value
float f; // float to hold received value
String text; // string to hold received text
VWComm vwc; // create an object to use with VWComm

void setup() {
  vw_set_rx_pin(receive_pin);
  vw_setup(2000); // Bits per sec
  vw_rx_start(); // Start the receiver
  Serial.begin(9600); // Initialise serial output
  Serial.println("Starting...");

}

void loop() {
  dt = vwc.dataType();
  if (dt == "B") {
    b = vwc.readByte();
    Serial.print("Byte received: ");
    Serial.println(b);
  }
  else if (dt == "I") {
    i = vwc.readInt();
    Serial.print("Integer received: ");
    Serial.println(i);
  }
  else if (dt == "S") {
    text = vwc.readStr();
    Serial.print("String received: ");
    Serial.println(text);
  }
  else if (dt == "F") {
    f = vwc.readFloat();
    Serial.print("Float received: ");
    Serial.println(f, DEC);
  }
```

```
  else {
    Serial.println("Data type not recognised!");
  }
}
```

The maximum distance will be indicated by frequent messages of "Data type not recognised!). We suggest that you use the table below to record your reliable transmission distances (***voltages are for the Tx module only***):

Baud Rate	Range 3.3V	Range 5V	Range 9V	Range 12V
2000 bps				
4000 bps				
6000 bps				

Our experiments suggest that the most significant factor is the Tx voltage. The baud rate appears to make little difference in this situation. While we have extended the Tx voltage as far as 12V, it may be more convenient to use 5V directly from the Arduino. Our measurement show that at 5V, the Tx draws a current of about 1 mA when transmitting and nothing at other times. Note that you may find some difficulty in achieving repeatable results

Stage 2 – Adding a Simple Straight Wire Aerial

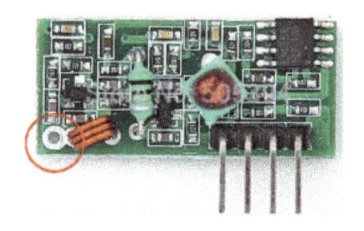

For this experiment and other stages, it will be useful to solder a female header onto Tx and Rx boards in the available aerial slots so that you easily plug in different aerials for comparison.

The calculations suggest an aerial length of 173mm. However, some factors can affect the optimum aerial length:

- *The modules already have a small helical aerial on the board.*
- *Electric current travels at approximately 5% less than the speed of light in metal.*
- *Wire resistance is affected by the diameter and material in the wire. RF flows in the skin of the copper wire, so the bigger the surface area, the smaller the resistive losses.*

Therefore, it is worth experimenting with various wire lengths between 160mm and 175mm to find the most effective aerial length.

Having added your straight wire aerials to the Tx and Rx, use the sketches in stage 1 to discover the best transmission distances with this type of aerial. For this stage of the experiment, we want to have 'line of sight' communication without any intervening obstacles – we shall consider these in stage 4.

One of the disadvantages of the straight-wire aerial is that it increases the overall sizes of the transmitter and receiver and this may be undesirable for some applications. Also, a protruding stiff wire could represent a physical danger to eyes – so beware!

We suggest that you use the tables below to record your results:

Quarter Wavelength:

Baud Rate	Range 3.3V	Range 5V	Range 9V	Range 12V
2000 bps				

Half Wavelength:

Baud Rate	Range 3.3V	Range 5V	Range 9V	Range 12V
2000 bps				

Stage 3 – Creating a Coil-Loaded Aerial

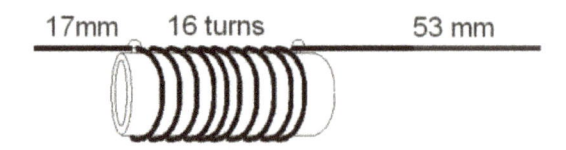

17mm   16 turns   53 mm

Using 25cm of 0.6mm insulated wire, a type of aerial called a *coil-loaded aerial (or antenna)* can be made. 17mm along the wire make 16 turns around a 2.5 mm cylinder. This should leave about 53mm of uncoiled wire. When fitted to the receiver this antenna should increase the reception range to around 25m line of sight and, within the house, increase reception through walls (see stage 4 for further details). Note that the material of the cylinder such as some plastics and loaded paper may affect the performance by a few percent.

Your results can be recorded in the table below:

Baud Rate	Range 3.3V	Range 5V	Range 9V	Range 12V
2000 bps				

Stage 4 – Factors that Affect Reception

We have mentioned that UHF transmission is limited to 'line of sight' communication. This basically means what it says and yet it is possible for signals to travel through such obstacles as walls, windows and human beings although there will be a loss of signal strength.

Atmospheric conditions can also affect transmission distances. The air always contains moisture in some form e.g. as vapour or in liquid and solid forms such as rain, mist, fog, hail, and snow. This normally attenuates the signal and has a greater effect on high frequency signals particularly those above 100 MHz. However, occasional conditions can give rise to a phenomenon known as *temperature inversion* where high frequency signals can travel much longer distances than normal.

See **http://www.tpub.com/neets/book10/40j.htm** for more details.

Another problem is that of electrical interference that may arise from other transmitters, electrical devices, and storms. The effect can be readily demonstrated by adding a second transmitter and observing received values. As many common devices use the 433 MHz frequency, this is a possible problem for remote control projects e.g we don't want next door's remote doorbell to switch our central heating on and off! Some such problems can be avoided by sending a special code, although the combined use of Virtual Wire and VWComm should ensure that only the intended signals are acted upon. Other sources of interference may be sporadic e.g. the automatic switching on and off of electrical devices or electrical storms. Fluorescent tubes can give rise to local interference and reception can be influenced by conducting objects with a similar wavelength.

The Experiments

4.1   *Obstacles*

In stages 2 and 3, we investigated the realistic maximum transmission range using aerials in an open space. These should be repeated with intervening objects such as walls, windows and human beings. Make a note of your results so that you can design remote control projects for use in various environments

4.2   *Atmospheric Conditions*

You will need to 'weatherproof' your Tx station for this part of the experiment. A simple piece of cling-film might suffice or you may wish to build a watertight box. As a first test, check what effect, if any, has been made by encasing your transmitter in a waterproof casing. Then you will need to place the waterproof Tx station outside and wait for various weather conditions. Remember that the relative humidity varies continually, so even some tests in 'dry' weather may yield different results.

4.3 *Electrical Interference*

While many of these effects can be difficult to reproduce and test, there are a few basic experiments that can illustrate the problems. In the first case, if we take possible devices transmitting on the same frequency, try placing the Tx near likely sources such as a computer or a remote-doorbell. Note that bluetooth devices operate at around 2.45 GHz and sufficiently above the frequency range to interfere with the 433 MHz modules. If you have a relay module, you could wire it in 'buzzer mode' to rapidly make and un-break. This will cause a magnetic field around the relay which might affect a nearby transmitter. Finally, you can try placing the transmitter near home appliances such as fridges and freezers, as well as fluorescent lights to see whether they generate any adverse effects.

Commercial Aerial Options

If you don't wish to make your own aerials, you could consider commercially available ones, although you will still need to make some physical connections to the Tx and Rx boards. The main options are:

Straight Wire (Whip) Aerials

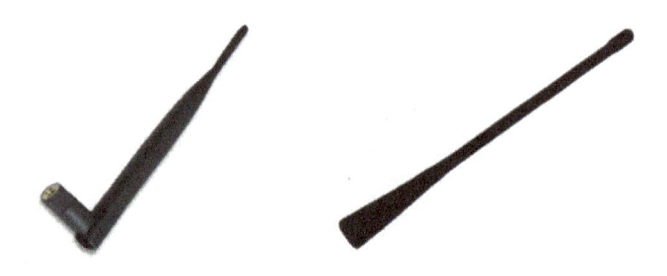

These aerials typically cost around £4 upwards and are widely available – search for *'whip antenna 433 MHz'*. Many are mainly aimed at commercial applications and require a connector to the transmitter.

Spiral Spring Helical Aerials

These very compact aerials consist of 5mm wire coiled for its whole length and can be soldered directly onto the boards. They are available in packs of 10 at a cost of around £2. They work well in extending the range to 20-30 metres but must be mounted vertically. Suitable for all the projects described in this book, we recommend these as an affordable and viable option.

Other varieties are available at the higher price of around £7, e.g. the flexible, 50 ohm, helical coil shown below:

Directional Aerials

In some cases, it may be necessary to detect only signals transmitted from a certain direction. For this purpose, a directional aerial, such as the Yagi, on the receiver is necessary. Yagi Aerials also have a greater range than the types described in this chapter but are unlikely to be needed for the projects described in this book.

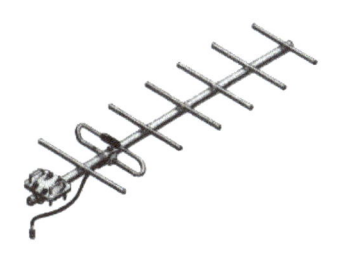

These aerials are typically used for television reception, which usually operate just above 470 MHz. Some are designed for the 433 Mhz range e.g. the LPRS Yagi 434A 7 element antenna, available for under £50.

What You Have Achieved and the Next Steps

In this experiment we have tested the range of the 433 MHz Tx/Rx modules in the context of practical home remote control projects. We have suggested that the optimum baud rate is 2000 bps and that a coiled wire antenna is probably the best option with a range of up to 30 metres. The transmission distance is essentially 'line of sight' and will be reduced by obstacles such as walls. The reliability of received data may or may not be of importance, depending upon the application. For example, a remote weather station might suffer a momentary break and transmit false or unintelligible signals and this may be of no significance. On the other hand, a remotely controlled door lock could present a serious challenge if the signals are not correctly sent and received!

When you are developing your own remote control/sensing application, you should bear in mind the issues that we have discussed and try to ensure that your system is sufficiently robust for its intended purpose. This includes software precautions as well as the hardware capabilities and should allow for spurious signals; the system must be able to recognise an error, deal with it, and permit a resolution. If it is possible to signal a problem to the user with an indication of what the problem is and what they should do, then so much the better!

Experiment 6: Making a Wireless Doorbell

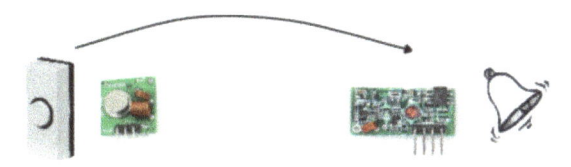

In this experiment we shall create a wireless doorbell, first as a simulation, then as a working project. We consider the various options available as well as practical issues. Finally, we look at modifying an existing doorbell for wireless operation.

Remote doorbells are commonplace these days and available fairly cheaply. However, they can have their problems in a close neighbourhood where they are all transmitting on the same frequency! Here, we shall create a basic remote control doorbell and then extend the design by considering various ways in which it might be improved.

Having established that we can realistically simulate a standard doorbell, we go on to look at practical problems and how we can exploit aspects of 'artificial intelligence' in the design of an advanced remote doorbell.

Consideration of Technical Issues

We start by emulating the actions of a standard wired doorbell in which the caller presses a button which causes a bell to ring for as long as the button remains pushed. For a remote doorbell, we can devise equivalent circuitry that will effectively duplicate this action by sending an 'ON' signal when a button is pressed at the transmitter end. In our simulation, the receiver will send a tone to a buzzer component each time an 'ON' signal is received and this tone will last for 1.5 seconds.

The first practical issue is that of range – a remote doorbell can be useful if you can take the receiver into the garden with you! This means that we want a range of, say, 30 metres and that will mean that you will have to attach suitable aerials to your Tx and Rx modules (see chapter 5 for details).

We then consider if it is necessary for the caller to hold the button pushed (modern systems don't require this) and what happens in this case with our system. Also what happens if the user keeps pressing the bell-push in an on/off manner?

In our introduction, we mentioned the possible problem of received signals from other wireless devices. Commercial wireless doorbells use a frequency between 300-433 MHz which is a range used by a range of other devices. While it may be possible to change channels (frequencies) on these devices, we are using a frequency fixed at 433.92 MHz, so we need an alternative strategy. Our bell-push signal involves sending a byte value of 1 using the VWComm library and since this requires the sending of a 'B' and a 1, it is unlikely that we shall receive spurious signals. However, if we have built several wireless controllers in our house, we could always change the 1 to another value in the range 0-255 and establish a unique code for our doorbell.

Also, we would like the user to have some control over the output that they can adjust the volume, ring tones and perhaps include a 'mute' facility.

We conclude the experiments by considering how we can take an existing standard doorbell and introduce modern electronics to transform it into a remotely controlled wireless doorbell.

The Experiments

Stage 1 – A Basic Doorbell Simulator

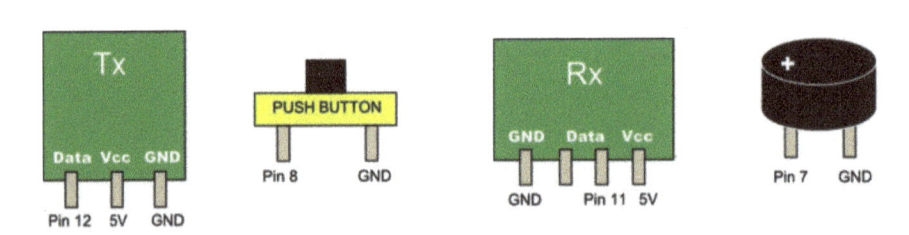

This is a very simple project to establish the basic principles and to form a building block for more sophisticated designs. The sketches are based upon the ones that we wrote when looking at simple remote switching in experiment 2. The connections for Tx and Rx are as shown above (note that some buzzers have a polarity indication).

Sketch Details

The transmitting station features a push-button that connects pin 8 to ground when it is pressed. Having set pin 8 as an input, we employ the internal pull-up resistor by writing a HIGH value to it, so that it is normally HIGH when unconnected and goes LOW when we push the button. The 'loop' section repeatedly checks the state of this pin and when it goes LOW, a byte value of 1 is transmitted.

The receiving station features a passive buzzer that represents the doorbell. When a byte value of 1 is received the sketch calls a standard Arduino function - tone – that outputs a sound representing the doorbell. The tone function has 3 parameters: pin, frequency (in Hz) and duration (in milliseconds), so in the receiving program we use tone(buzzer,2000,1500) which outputs a tone of 2 KHz for 1.5 seconds to pin 7 (named *buzzer*).

Sketch Code

The Transmitter Code is as follows:

```
/*
  Expt6_Stage1_Tx - Transmitting Sketch
  Basic Doorbell Simulator
*/

#include <VirtualWire.h>
#include <VWComm.h>
const int transmit_pin = 12; // Tx data to D12
const int inPin = 8; // push button to D8
VWComm vwc; // create an object to use with VWComm

void setup() {
  vw_set_tx_pin(transmit_pin);
  vw_setup(2000); // Bits per sec
  pinMode(inPin, INPUT); // set inPin as an input
  digitalWrite(inPin, HIGH); // use pull-up
```

```
  Serial.begin(9600); // initialise serial output
}

void loop() {
  if (digitalRead(inPin) == LOW) { //check for btn push
    vwc.sendByte(1); // send 'press' value
    Serial.println("Sending data");
    delay(300); // short delay
  }
}
```

The **Receiver Code** is as follows:

```
/*
  Expt6_Stage1_Rx - Receiving Sketch
  Basic Doorbell Simulator
*/

#include <VirtualWire.h>
#include <VWComm.h>

const int receive_pin = 11; // Rx data to D11
const int buzzer = 7; // Buzzer anode to D7
String dt; // string to received data type
byte b; // byte to hold received value
VWComm vwc; // create an object to use with VWComm

void setup()
{
  vw_set_rx_pin(receive_pin);
  vw_setup(2000);    // bits per sec
  vw_rx_start();     // start the receiver running
  pinMode(buzzer, OUTPUT); // set pin as an output
}

void loop()
{
  dt = vwc.dataType(); // wait for incoming data
  if (dt == "B") { // code 'B' indicates 'byte'
    b = vwc.readByte(); // get the byte of data
    if (b == 1) {
      tone(buzzer, 2000, 1500);
    }
  }
}
```

Stage 2 – Possible Problems and How to Overcome Them

As we mentioned earlier, one of the biggest problems with remote doorbells is receiving spurious signals from other remote devices located nearby and we have concluded that our projects are not likely to be affected by such problems. However, we might wish to deal with situations where the caller keeps the push-button pressed for a long time, or when the push-button is repeatedly pressed in an ON/OFF fashion. On one hand, we do not wish the doorbell to sound more than once if we have heard it, but on the other, we would like the caller to repeat the signal after a short delay in case we missed the first call.

A traditional doorbell typically uses an electromechanical solenoid which is wired to break its own circuit when power is applied, but to then restore the connection when the contact returns to its 'normally closed' position by a spring. The basic circuit is shown below:

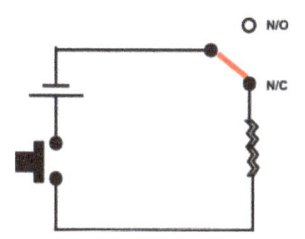

A simple version of this can be made using a relay module. Actual doorbells combine the contact (shown in red above) with a piece of metal that bangs against a bell (imagine the 'normally-open' contact in the diagram to be a bell). The doorbell will continue to ring while the button is held and the caller can opt for a short ring, long ring, or multiple rings.

Sketch Details

Our doorbell is different in that the bell push sends an 'ON' signal to the receiver which causes the 'bell' to sound for a short time as given by the third parameter in tone(pin,frequency,duration). The transmitter will repeat its 'ON' signal while the button remains pressed with a short delay between signals so the result will be similar to a traditional doorbell.

The problems of continual pressing or repeated pressing are perhaps best dealt with at the Tx end, but it is necessary to first define exactly what we want to happen. Let us start with the continual bell push. How do we detect this and how do we limit the 'ON' signals being transmitted? Fortunately, there is a simple solution which is to include a delay in the transmitter code after an 'ON' signal has been sent. This is a matter of individual choice, but a 5 second delay might be a reasonable option.

With regard to multiple presses of the button, the inclusion of a delay (as suggested above) should also serve to solve the problem. An additional piece of code could be included to ignore further presses in a given period, although this is probably not worthwhile.

If you wish to include a 'mute' facility, this should be done at the receiver end. This can be easily achieved with a simple push button that toggles between 'sound-on' and 'sound-off'. It would be useful to indicate the state with a green LED that lights when the audio is active.

Sketch Code

The Transmitter Code should include a delay to limit constant or repeated bell pushes. Change the short delay at the end of the loop to:

```
delay(5000);
```

The Receiver Code will include a 'mute' facility. We create a boolean variable, 'mute' which we initially set to 'false' (meaning 'allow sound'). We want to set this to 'false' or 'true' whenever we press a button but here we encounter a difficulty because the sketch is set to wait for an incoming data signal. The simplest way around this is to use a mechanism known as an *interrupt* which, in our case is a little routine that is activated when we set a digital pin to LOW when the mute button is pressed. The 'mute' variable is toggles between TRUE and FALSE. The additional circuitry is as shown:

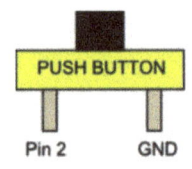

The simulation incorporates an LED that lights when the button is pressed, remaining ON for a short time. This acts as a visual indicator which is useful when the sound is muted. For simplicity, this experiment uses the on-board LED attached to pin 13.

The <u>Receiver Code</u> becomes:

```
/*
  Expt6_Stage2_Rx - Receiving Sketch
  Enhanced Doorbell Simulator using Interrupt (pin 2)
*/

#include <VirtualWire.h>
#include <VWComm.h>

const int receive_pin = 11; // Rx data to D11
const int buzzer = 7; // buzzer anode to D7
const int ledPin = 13; // LED anode to D13
volatile boolean mute = false; // use with interrupt
String dt; // string to received data type
byte b; // byte to hold received value
VWComm vwc; // create an object to use with VWComm

void setup()
{
  vw_set_rx_pin(receive_pin);
  vw_setup(2000);   // Bits per sec
  vw_rx_start();    // Start the receiver running
  pinMode(buzzer, OUTPUT); // buzzer pin as output
  pinMode(ledPin, OUTPUT); // LED pin as output
  digitalWrite(ledPin, LOW); //start with LED off
  attachInterrupt(digitalPinToInterrupt(2), toggleMute, FALLING);
  digitalWrite(2, HIGH); // set interrupt pin HIGH
}

void loop()
{
  dt = vwc.dataType(); // wait for incoming data
  if (dt == "B") { // code 'B' indicates type 'byte'
    b = vwc.readByte(); // get the byte of data
    if ((b == 1) && (!mute)) {
      soundBuzzer();
    }
  }
}
```

```
void soundBuzzer()
{
  digitalWrite(ledPin, HIGH); // switch on led
  tone(buzzer, 2000, 2000); // sound buzzer
  delay(3000);
  digitalWrite(ledPin, LOW); // switch off led
}

void toggleMute()
{
  mute = !mute; // toggle mute
  delay(1000); // allow time to release button
}
```

Further details about interrupts can be found in 'Exploring Arduino' by Jeremy Blum and also in the reference section of the Arduino website.

Stage 3 – A Practical Doorbell: Construction Notes and Components

This stage involves making a working wireless doorbell with the minimum of components. Essentially it is a development of the previous stage with some practical construction details. The following additional components are required:

> Standard bell-push
> Loud buzzer module
> Batteries (and connectors) or PSU
> Suitable containers to house Tx and Rx

Construction Notes

We have chosen to use a 'loud buzzer' module which produces a louder tone than seems to be available from the buzzer component by itself. It is inexpensive and easy to use. The buzzer module has 3 connections: GND, I/O, and VCC. These will be connected to GND, Digital Pin 7, and 5V respectively on the Arduino.

The additional connections previously described still apply.

Sketch Code

The previous code examples can still be used for this stage, however the following code might be more suitable with the buzzer module:

```
void soundBuzzer()
{
  digitalWrite(ledPin, HIGH); // switch on led
  tone(buzzer,2000);
  delay(1000);
  tone(buzzer,1000);
  delay(1000);
  digitalWrite(buzzer,LOW);
  digitalWrite(ledPin, LOW); // switch off led
}
```

Stage 4 – Modifying an Existing Doorbell

Modifying an existing wired doorbell into a wireless version is straightforward and relatively inexpensive. The existing bell-push can be used and it is suggested that it is wired to a transmitter unit inside the door using the current wiring as far as is necessary. This unit will also require a power supply that can be a battery or a mains powered PSU. In either case, the unit needs to be situated so as to be convenient for replacing the battery or access to a mains socket. It can be quite small if we use an Arduino Nano (or micro) and will require a small aerial.

The receiving unit will be used to replicate the actions of the bell-push and this may vary according to the type of doorbell being used. Older types require the caller to hold the bell-push for a short time while later 'ding-dong' versions only require a momentary bell-push. Both types use a solenoid based mechanism and assuming that the power supply (usually batteries) is still available at the bell, the receiver only needs to replicate the action of the bell push. This can be achieved with a simple relay module, whereby the contacts will be held closed for a short time. This time will need to be longer if the bell is of the make-break type (as opposed to the ding-dong chime variety). Our experiment uses a 2 second delay.

Construction Notes

The transmitter will send a simple ON signal as in stage 1 and the bell push wires will be connected to pin 8 and GND of the transmitter. The receiver is also simple in that upon receiving the 'ON' signal, it will send a signal to the relay and hold it for a short time.

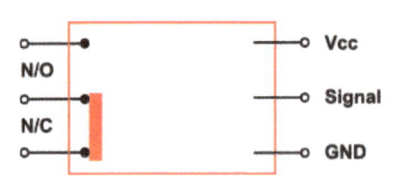

Connect the wires that used to come from the bell push to the normally open (N/O) connections on the relay module. Connect Vcc to +5V on your arduino and GND to GND. Finally, connect 'Signal' to a convenient digital pin (we use pin 7 in our example).

Sketch Code

For the transmitter, use the same code as in stage 1.

The Receiver Code will be as follows:

```
/*
  Expt6_Stage4_Rx - Receiving Sketch
  Modifying an Existing Doorbell
*/

#include <VirtualWire.h>
#include <VWComm.h>
const int receive_pin = 11; // Rx data to D11
const int doorbell = 7; // relay connection to D7
String dt; // string to received data type
byte b; // byte to hold received value
VWComm vwc; // create an object to use with VWComm

void setup()
{
  vw_set_rx_pin(receive_pin);
  vw_setup(2000); // Bits per sec
  vw_rx_start(); // Start the receiver running
```

```
  pinMode(doorbell, OUTPUT); // set pin as output
  digitalWrite(doorbell, LOW); // ensure doorbell is OFF
}
void loop()
{
  dt = vwc.dataType(); // wait for incoming data
  if (dt == "B") { // code 'B' indicates 'byte'
    b = vwc.readByte(); // get the byte of data
    if (b == 1) {
      digitalWrite(doorbell, HIGH); // sound doorbell
      delay(2000);
      digitalWrite(doorbell, LOW); // doorbell OFF
    }
  }
}
```

What you have Achieved and the Next Steps

In this experiment you have used the 433 modules to develop a practical wireless doorbell and considered some practical issues that could influence your design. The stages of this experiment have taken us through the various considerations involved in developing a practical remote-control application and how we might adapt existing facilities. It is hoped that this should be useful guidance for developing applications in other practical situations.

This experiment has concentrated on what we can do at the receiver end, but we could employ some broad thinking to the doorbell design and develop a project that involves options for the caller. For example, we could extend the simple bell-push to a numeric keypad where friends could enter individual codes only known to themselves, but casual callers would be required to enter another code. The receiver could examine an incoming call and display information on an LCD display. We could also consider the option of adding a response facility to remotely transmit a message back to the caller (although there is the possibility of a conflict between the two transmitters operating on the same frequency).

Experiment 7: Simple Weather Station

A weather station provides regular data values relating to prevailing weather conditions using various sensors. Typical data includes temperature, atmospheric pressure and relative humidity, although some stations also include rain gauges and wind monitors (speed and direction). It is reasonably simple to create a weather station using readily available and inexpensive sensors which output their data to an Arduino for processing.

In this experiment, we shall create a simple weather station that monitors temperature, pressure and relative humidity. This would typically be located outdoors and use a 433 Tx module to send data values to an indoor receiver which, in turn, displays the data on an LCD screen together with the date and time.

The weather station is built up in a series of stages in which the main components are introduced one-by-one, so that their functions can be examined and verified. The experiment uses the essential Tx and Rx programs introduced in the previous chapters with the VWComm library.

Data is obtained using two readily available and inexpensive sensors: the DHT11 and the BMP180. A link to datasheets for these can be found on the website accompanying this book (see also Bibliography).

Consideration of Technical Issues

We are going to monitor temperature, atmospheric pressure and relative humidity. While we can often obtain these values to numerous decimal places, we should consider how much precision that we require and the accuracy of the sensors. For example, it is meaningless to display a temperature to 2 decimal places if the sensor accuracy is +/- 1 degree. The relevant details will be given as each sensor is presented in the particular stage of the experiment, but the following describes the technical background regarding the measured quantities.

Temperature is measured in degrees Celsius (°C). This is measured directly from the DHT11 sensor (see stage 1 below) to an accuracy of +/- 2°C. If you wish to display the temperature in degrees Fahrenheit (°F), use the conversion: F = 9C/5 + 32.

Atmospheric Pressure is traditionally measured in bars (or millibars). The measuring device is called a barometer, giving rise to the term *'barometric pressure'*. It is a good indicator of changing weather conditions e.g. a low pressure reading is associated with rain, while a high pressure suggests sunshine. The direction of change (e.g. high to low) can be a good prediction of future weather conditions. The atmospheric pressure diminishes with height and also changes with temperature. We can also use readings of atmospheric pressure to determine actual height above ground level for a given location. Ground level is, itself, measured in terms of sea level, so any calculations will be based upon the pressure at sea level, the pressure at ground level, and the pressure at the point of measurement. The sketch code in this experiment uses standard formulae for which there are various sources (see Bibliography). When using the example code, you will need to modify the relevant line to refer to your height above sea level in metres. The atmospheric pressure is obtained from the BMP180 sensor which has an absolute accuracy of approximately +/- 1 millibar. Alternative measures are: inches of mercury ("Hg), atmospheres (atm), kiloPascals (kPa). They are related according to: 1 atm = 29.92 "Hg = 101.325 kPa = 1013.25 mb. We also often refer to hPa which equals 100 Pa that gives a unit equivalent to the millibar.

Relative Humidity is the actual vapour density in the air compared to the saturation vapour density (i.e. what the air can hold at that temperature). Vapour density is usually measured in grammes per cubic metre and relative humidity is usually expressed as a percentage. It can be measured using the DHT11 sensor which provides data values in percentages to an accuracy of +/- 5% RH.

This stages of this experiment can be conducted indoors and without aerials. However, a realistic weather station will involve situating the transmitter and its sensors outside and protecting it from the elements. Also, the transmitter and receiver will need aerials to provide an adequate range. See the earlier chapter (Experiment 5) for more details.

The Experiments

Stage 1 – Using the DHT11 to send humidity and temperature readings

Introducing the DHT11

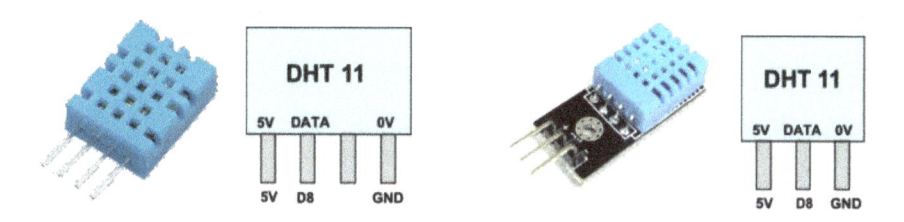

The DHT11 sensor is a relatively cheap sensor that measures both temperature and humidity. The DHT21 and DHT22 sensors can also be used but these sensors use a different library. The DHT11 is available in two forms: simple sensor (4 pins), or as a module (3 pins), sometimes referred to as an 'electronic brick'. If using the simple sensor, you should also use a pullup resistor (4.7KΩ to 10KΩ). The module has a built in pullup resistor.

Only three pins on the DHT11 need be connected. These are 0V, +5v and a single data line. The DHT library enables both temperature and humidity values to be read from a single data pin. The humidity range measured is between 20-90% with an accuracy of +/- 5%. The data pin should be connected to a digital input.

Further details are available online (see Bibliography)

Sketch Details

Connect the DHT11 data pin to D8 on the Arduino and connect the 5V and 0V to 5V and GND.

The Tx Sketch reads humidity and temperature values from the DHT11 and sends text and data values using VWComm. The maximum sampling rate for the DHT11 is once per second, so we need to incorporate a short delay in our loop.

Rx Sketch – uses a slightly modified template Rx to report values on serial monitor. Not all of this code is used. Note that the Rx waits for a 'start' signal from the Tx, so this should be started before the Tx.

Sketch Code

The <u>Transmitter Code</u> is as follows:

```
/*
  Expt7_Stage1_Tx - Transmitting Sketch
  Humidity and Temperature values using DHT11
*/

#include <VWComm.h>
#include <dht.h>
#include <VirtualWire.h>
const int transmit_pin = 12; // Tx data to D12
const int transmit_en_pin = 3; // reset default
const int dht_dpin = 8; // DHT11 data pin to D8
int x; // variable to hold data values
dht DHT; // create an object to use with dht library
VWComm vwc; // create an object to use with VWComm

void setup()
{
  vw_set_tx_pin(transmit_pin);
  vw_set_ptt_pin(transmit_en_pin); // reset default
  vw_setup(2000); // bits per sec
  vwc.sendStr("Start"); // synchronise with Rx
  delay(500); // delay to allow Rx to synchronise
}

void loop()
{
```

```
  DHT.read11(dht_dpin); // get data values
  x = int(DHT.humidity); // humidity component
  vwc.sendStr("Humidity:"); // send text
  vwc.sendInt(x); // send value
  x = int(DHT.temperature); // temperature component
  vwc.sendStr("Temperature:"); // send text
  vwc.sendInt(x); // send value
  delay(2000); // Pause for 2 seconds.
}
```

The <u>Receiver Code</u> is as follows:

```
/*
  Expt7_Stage1_Rx - Receiving Sketch (standard)
  Receive and Display Humidity and Temperature values
*/
#include <VWComm.h>
#include <VirtualWire.h>
const int receive_pin=11; // Rx data to D11
String text; // string to hold received text
int i; // integer to hold received values
byte b; // byte to hold received values
float f; // float to hold received values
String dt; // string to hold received data type
VWComm vwc; // create an object to use with VWComm

void setup()
{
  vw_set_rx_pin(receive_pin);
  vw_setup(2000); // Bits per sec
  vw_rx_start(); // Start the receiver
  Serial.begin(9600); // initialise serial output
  Serial.println("Waiting for start command");
  do
  {
    dt = vwc.dataType();
    if (dt == "S") {
      text = vwc.readStr();
    }
  } while (text != "Start");
  Serial.println("Receiving data ....");
}

void loop()
{
  dt = vwc.dataType();
  if (dt == "B") {
```

```
    b = vwc.readByte();
    Serial.print("Byte received: ");
    Serial.println(b);
  }
  else if (dt == "I") {
    i = vwc.readInt();
    // Serial.print("Integer received:");
    Serial.println(i);
  }
  else if (dt == "S") {
    text = vwc.readStr();
    // Serial.print("String received:");
    Serial.print(text);
  }
  else if (dt == "F") {
    f = vwc.readFloat();
    Serial.print("Float received: ");
    Serial.println(f, DEC);
  }
  else {
    Serial.println("Data type not recognised!");
  }
}
```

Stage 2 – Using the BMP180 to send pressure readings

Introducing the BMP180

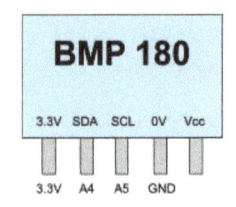

The BMP180 is a low-cost (approx. £4) sensor that measures barometric pressure (it also measures temperature, but we already have that from the DHT11). Although the sensor operates on 3.3V, most boards feature a regulator that enables connections to a 5V supply via the Vcc pin. The sensor is connected via the I2C bus that features the SDA and SCL signals that connect to pins A4 and A5 respectively on the Arduino Uno. The sensor has level shifters to ensure that the voltage levels on these pins are also compatible. While devices on the I2C bus normally require pull-up resistors (typically 4.7 kΩ), the BMP180 incorporates these in the module so that all connections are straight to the Arduino with no extra circuitry.

Most BMP180 modules are supplied with a 5 pin header that needs to be soldered onto the module.

Sketch Notes

The BMP 180 measures the atmospheric pressure in Pascals (Pa) at the current location. To enable comparisons between different locations, we normally standardise these to sea-level. The average pressure at sea-level is 1 atmosphere (atm) = 1013.25 mb.

There are different libraries available to communicate with the BMP180. We have chosen to use the SFE_BMP180 library available from:

> https://github.com/sparkfun/BMP180_Breakout_Arduino_Library/archive/master.zip

To include this library in your sketch, use:

```
#include <SFE_BMP180.h>
```

To create an object called *pressure* from which we can obtain atmospheric pressure and altitude measurements, we use the following code in the declarations before the setup section:

```
SFE_BMP180 pressure;
```

The Rx sketch uses the basic template receiver code included in the VWComm library to report values on serial monitor. The code on the accompanying website contains minor variations to improve display.

Sketch Code

Thanks to Mike Grusin, SparkFun Electronics, for the example sketch that comes with the library, on which the following code is based:

The **Transmitter Code** is as follows:

```
/*
  Expt7_Stage2_Tx - Transmitting Sketch
  Pressure and altitude values using BMP180
*/

#include <VirtualWire.h>
#include <VWComm.h>
#include <SFE_BMP180.h>
#include <Wire.h>

#define ALTITUDE 100.0 // Coventry altitude (metres)

double P; // storage for pressure readings
const int transmit_pin = 12; // Tx data to D12
const int transmit_en_pin = 3; // reset default

SFE_BMP180 pressure; // create an object for BMP values
VWComm vwc; // create an object to use with VWComm

void setup()
{
  Wire.begin(); // create a Wire object
  vw_set_tx_pin(transmit_pin); // reset default
  vw_set_ptt_pin(transmit_en_pin); // reset default
  vw_setup(2000); // Bits per sec
  vwc.sendStr("Start"); // synchronise with Rx
  delay(200); // delay to allow Rx to synchronise
  if (pressure.begin()) // initialise BMP180
    vwc.sendStr("BMP180 init success");
  else
  {
    vwc.sendStr("BMP180 init fail");
    while (1); // Pause forever.
  }
}

void loop()
{
  P = getPressure(); // get a new pressure reading:
  vwc.sendStr("Pressure (mb):"); // send text
  delay(500);
  vwc.sendFloat(float(P), 4); // send pressure value
  delay(2000);
}
```

```
double getPressure() // function to read pressure
{
  char status;
  double T, P, p0;
  status = pressure.startTemperature();
  if (status != 0)
  {
    delay(status);
    status = pressure.getTemperature(T);
    if (status != 0)
    {
      status = pressure.startPressure(3);
      if (status != 0)
      {
        delay(status);
        status = pressure.getPressure(P, T);
        if (status != 0)
        {
          p0 = pressure.sealevel(P, ALTITUDE);
          return (p0);
        }
      }
    }
  }
}
```

For a fully commented version, refer to the example with the library (download details given on page 83).

The receiver code is as before.

Stage 3 – Adding a Real-Time Clock (RTC)

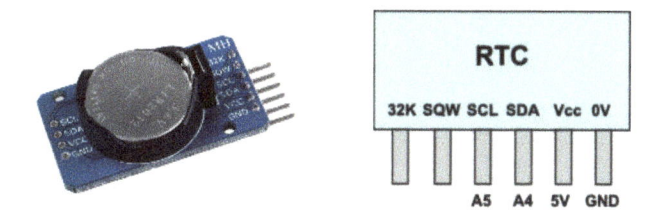

The Real-Time Clock module is connected via the I2C bus with pin SDA to A4 and SCL to pin A5 on the Arduino Uno. Only four pins are connected as shown below – ignore the other pins. The modules have duplicate connections to permit the use of other devices on the same bus and we could use these to connect the BMP 180. However, the RTC could be situated at either the Tx end or the Rx end and it makes things simpler to incorporate this into the receiver. For now, let us just consider the basic RTC code and we will put everything together in Stage 4.

Sketch Details

The sketch is quite simple and mainly self-explanatory. The Wire library is included to provide communications via the I2C bus and the RTClib library allows simple sketch commands to reset or access times and dates. We create an object called rtc, using the statement: RTC_DS1307 rtc; and we can then use its associated method rtc.adjust and property rtc.now. Note that rtc.adjust sets the time and date using the values from your computer at the time of compilation. The skecth outputs date and time to the serial monitor. There is no receiver sketch.

Sketch Code

```
/*
   Expt7_Stage3 Stand-Alone Sketch
   Display date and time values from RTC
*/
#include <Wire.h>
#include "RTClib.h"
RTC_DS1307 rtc; // create an object to use with RTC

void setup () {
  Wire.begin(); // create a Wire object
  Serial.begin(9600); // initialise serial output
  rtc.adjust(DateTime(F(__DATE__), F(__TIME__)));
}

void loop () {
  DateTime now = rtc.now(); // get date and time
  Serial.print(now.day(), DEC); // output date values
  Serial.print('/');
  Serial.print(now.month(), DEC);
  Serial.print('/');
  Serial.print(now.year(), DEC);
  Serial.print(" ");
```

```
    Serial.print(now.hour(), DEC); // output time values
    Serial.print(':');
    Serial.print(now.minute(), DEC);
    Serial.print(':');
    Serial.print(now.second(), DEC);
    Serial.println();
    delay(3000); // delay 3 seconds
}
```

Further notes: the limitation of the above program is that it will reset the RTC every time it is run, but it will use the time of the original compilation! This can be avoided by changing the rtc.adjust line to:

```
if (! rtc.isrunning()) {
   rtc.adjust(DateTime(__DATE__, __TIME__));
}
```

An alternative to multiple Serial.print commands is to build up a string and use a single println. For example:

```
String thisdate = String(now.day());
thisdate = thisdate + "/";
thisdate = thisdate + String(now.month());
Serial.println(thisdate);
```

Note the use of single inverted commas in the second line. The display won't show leading zeros e.g. the 2nd January 2017 would appear as 2/1/2017 whereas we might wish a consistent 2-digit display for day and month. This can be achieved by checking the length of the particular string component e.g.

```
String thismonth = String(now.month());
if(thismonth.length()<2){
    thismonth = "0" + thismonth;
}
thisdate = thisdate + thismonth;
Serial.println(thisdate);
```

Stage 4 – Outputting values to an LCD display

In this simple weather station, we shall display the current readings for time, date, temperature, atmospheric pressure and relative humidity on an LCD display. While the 2-line, 16-character display is commonplace, we shall use the 4-line, 20 character display in order to display the full information available. These are not much more expensive (around £3) and have similar connections.

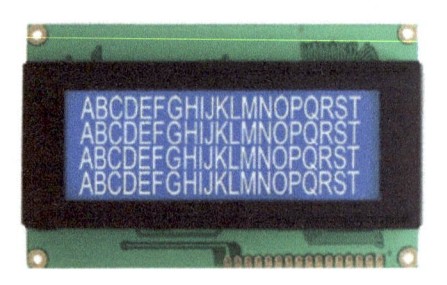

The following diagram shows how the display is connected to the Arduino. LCD pins 7,8,9, and 10 are not used in this application and pins 15 and 16 provide the optional backlight to the display. The 10K potentiometer adjusts the contrast of the display – if the display appears blank then try adjusting the potentiometer to reduce or increase contrast before checking all your connections!

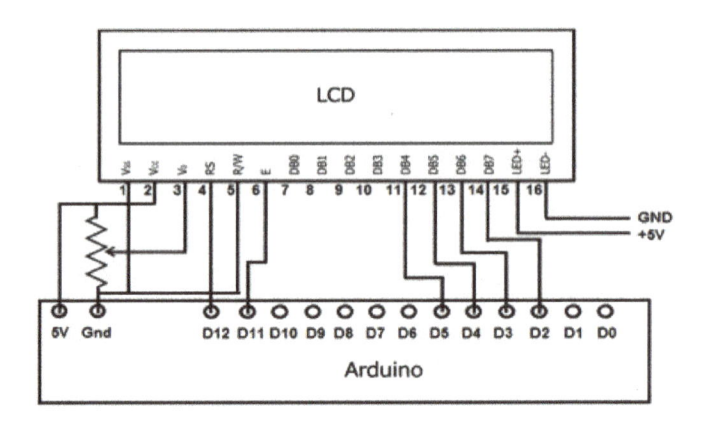

Sketch Details

In this experiment we have used an LCD display to present the values received from a remote sensing station. One disadvantage is that the connections take up most of the digital pins of the receiving Arduino, although it leaves sufficient for our requirements. Sketches can use simple commands using the *Liquid Crystal* library e.g. setup, clearing the display, positioning text etc. An alternative is to use an LCD shield that greatly simplifies the connections because they plug into the Arduino and usually incorporate some control buttons that can monitored from an analogue pin (see experiment 9) Some new varieties of the 4020 LCD display involve fewer connections because they use the I2C bus for which a different library will be required.

Sketch Code

The Transmitter Code is as follows:

```
/*
  Expt7_Stage4_Tx - Transmitting Sketch
  Pressure values using BMP180
  Temperature and humidity values from DHT11
*/

#include <VirtualWire.h>
#include <VWComm.h>
#include <SFE_BMP180.h>
#include <Wire.h>
#include <dht.h>
#define ALTITUDE 100.0 //Coventry altitude (metres)

double P; // storage for pressure readings
const int transmit_pin = 2; // Tx data to D2
const int transmit_en_pin = 3; // reset default
const int dht_dpin=8; // DHT11 data pin to D8
int hum; // variable to hold humidity values
int x; // variable to hold other values

SFE_BMP180 pressure; // create an object for BMP values
dht DHT; // create an object to use with DHT library
VWComm vwc; // create an object to use with VWComm

void setup()
{
  vw_set_tx_pin(transmit_pin); // define Tx data pin
  vw_set_ptt_pin(transmit_en_pin); // reset default
  vw_setup(2000); // Bits per sec
  vwc.sendStr("Start"); // synchronise with Rx
  delay(200); // delay to allow Rx to synchronise
  if (pressure.begin())
    vwc.sendStr("BMP180 init success");
  else
```

```
  {
    vwc.sendStr("BMP180 init fail");
    while (1); // Pause forever.
  }
}

void loop()
{
  P = getPressure(); // get a new pressure reading:
  vwc.sendStr("Pr (mb):"); // send text
  delay(200);
  vwc.sendFloat(float(P), 2); // send pressure value
  delay(200);
  DHT.read11(dht_dpin); // read data values
  x = int(DHT.humidity); // get humidity component
  vwc.sendStr("Hum % ="); // send text
  delay(200); // allow Rx to receive and report
  vwc.sendInt(x); // send humidity value
  delay(200); // allow Rx to receive and report
  x = int(DHT.temperature); // get temperature component
  vwc.sendStr("T.Celsius ="); // send text
  delay(200); // allow Rx to receive and report
  vwc.sendInt(x); // send temperature value
  delay(2000); // pause for 2 seconds
}

double getPressure()
{
  char status;
  double T, P, p0;
  status = pressure.startTemperature();
  if (status != 0)
  {
    delay(status);
    status = pressure.getTemperature(T);
    if (status != 0)
    {
      status = pressure.startPressure(3);
      if (status != 0)
      {
        delay(status);
        status = pressure.getPressure(P, T);
        if (status != 0)
        {
          p0 = pressure.sealevel(P, ALTITUDE);
          return (p0);

        }
```

```
      }
    }
  }
}
```

The sketch for the <u>Receiver</u> is given below. Note that in its present form the time is displayed as hours, minutes and seconds. The day and year data is also available but we have chosen not to display it.

```
/*
  Expt7_Stage4_Rx Receiver Sketch
  Display weather data and time values
*/

#include <Wire.h>
#include <RTClib.h>
#include <LiquidCrystal.h>
#include <VirtualWire.h>
#include <VWComm.h>

const int receive_pin = 7; // Rx data to D7
String text; // string to hold received text
int i; // integer to hold received values
byte b; // byte to hold received values
float f; // float to hold received values
String dt; // string to hold received data type
byte line = 0; // line indicator for lcd display
byte count = 0; // data count to refresh lcd display

VWComm vwc; // create an object to use with VWComm
RTC_DS1307 rtc; // create an object to use with RTC
// initialize library with interface pin details
LiquidCrystal lcd(12, 11, 5, 4, 3, 2);

void setup()
{
  Wire.begin(); // initialise Wire for use with clock
  rtc.begin(); // initialise real time clock
  if (! rtc.isrunning()) { // update clock
    rtc.adjust(DateTime(__DATE__, __TIME__));
  }
  vw_set_ptt_pin(13); // reset default
  vw_set_rx_pin(receive_pin); // reset default
  vw_setup(2000); // Bits per sec
  vw_rx_start(); // Start the receiver
  lcd.begin(20, 4); // initialise LCD display
}
```

```
void loop()
{
  dt = vwc.dataType(); // wait for and identify data
  if (dt == "B") {
    b = vwc.readByte(); // byte received
    lcd.setCursor(0, line);
    lcd.print(b);
  }
  else if (dt == "S") {
    text = vwc.readStr(); // string received
    if (count == 0) {
      lcd.clear();
    }
    lcd.setCursor(0, line);
    lcd.print(text);
  }
  else if (dt == "I") {
    i = vwc.readInt(); // integer received
    lcd.print(i);
  }
  else if (dt == "F") {
    f = vwc.readFloat(); // float received
    lcd.print(f);
  }
  count = count + 1; // count data values received
  line = count / 2; // set line on lcd display
  if (line > 2) { // reset display and show time
    line = 0;
    count = 0;
    lcd.setCursor(0, 4);
    DateTime now = rtc.now();
    lcd.print (" Time " );
    lcd.print(now.hour(), DEC);
    lcd.print(":");
    lcd.print(now.minute(), DEC);
    lcd.print(":");
    lcd.print(now.second(), DEC);
  }
}
```

What you have Achieved and the Next Steps

In this experiment, we have constructed a simple – yet practical – remote weather station. While we have explored some of the devices for measuring the main factors of temperature, relative humidity, and atmospheric pressure, we have noted the accuracy of the values and have reflected this in the displayed values. We have only used inexpensive sensors and the accuracy is only suitable for domestic purposes although greater accuracy could be obtained without significant extra outlay.

Additional features could include a rain gauge and wind monitor. We could also improve the display by using a TFT module such as the ones used in mobile phone. These are available quite cheaply e.g. £10 with various connection options, although they do require some modest programming skills to achieve a good display.

Experiment 8: Remote Keypad Lock

In this experiment, we shall transfer values from a matrix keypad to a 433 transmitter and use it to send a 4-digit security code to a receiver. We build up the experiment by first displaying the key values on the serial monitor and eventually create a simple receiver program that checks the incoming values against a stored code and sets LED lights accordingly. Finally, we consider how this program can be modified to act as a practical lock/unlock mechanism.

We shall use a standard 4x4 matrix keyboard for this experiment and these are readily available for around £2. The Receiver uses an RGB LED for the simulation. It is more convenient to use a dedicated RGB module when experimenting because they have built-in resistors and are inexpensive, about £1, if you shop around. For a practical implementation, you will require a relay module to power an electronic lock. Various 12V locks are available and these typically draw a current of 1A and so a relay module can be used to switch the lock from an Arduino.

Consideration of Technical Issues

The 4x4 keypad is based upon 4 horizontal wires and 4 vertical wires resulting in a matrix of interconnection points as shown overleaf.

From this, we can see that we have 8 lines connecting the keypad to the Arduino. A single keypress causes a connection to be made between a given row and column. The keypad library identifies the row and column and translates the result into an ASCII character code.

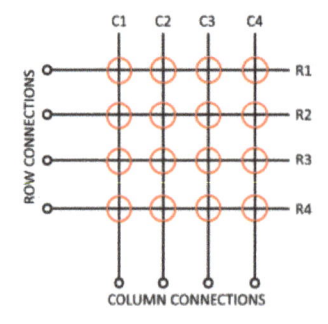

The connections are numbered 1-8 from left to right (viewing from the front of the keypad). 1-4 are the row connections and 5-8 are the column connections. These should be connected in the order 1-8 to digital pins 2-9 on the Arduino.

The data pin of the 433 MHz Tx will be connected to pin 12 of the Arduino. Also connect the Vcc and GND to 5V and 0V respectively.

The Experiments

Stage 1 – 4x4 Matrix Keyboard – Displaying on the Serial Monitor

This is a very simple experiment to demonstrate that we can read key presses from the keyboard. The transmitter will wait until a key is pressed and then send the corresponding ASCII code. For the moment, we do not include a receiver, but simply display the key values on the serial monitor.

Sketch Details

This sketch uses the Keypad library which is available at:

> http://playground.arduino.cc/Code/Keypad

along with details of how to use it. We have also incorporated the details for transmitting the key code as a byte of data using the VWComm library. These lines of code will be familiar to you by now.

Sketch Code

The Transmitter Code is as follows:

```c
/*
  Expt8_Stage1_Tx - Using digital keypad
  Output to serial monitor (no Rx for this stage)
*/

#include <Keypad.h>
#include <VirtualWire.h>
#include <VWComm.h>
const int transmit_pin = 12; // Tx data to D12
const byte ROWS = 4; // define number of rows
const byte COLS = 4; // define number of columns
char keys[ROWS][COLS] = { // create an array of chars
  {'1','2','3','A'},
  {'4','5','6','B'},
  {'7','8','9','C'},
  {'*','0','#','D'}
};
byte rowPins[ROWS] = {2,3,4,5}; // define row pinouts
byte colPins[COLS] = {6,7,8,9}; // define col pinouts
VWComm vwc; // create an object to use with VWComm
Keypad keypad = Keypad( makeKeymap(keys), rowPins,
        colPins, ROWS, COLS ); // create a keypad object

void setup(){
  vw_set_tx_pin(transmit_pin);
  vw_setup(2000);          // Bits per sec
  Serial.begin(9600); // initialise serial output
}

void loop(){
  char key = keypad.getKey(); // read keypad input
  if (key != NO_KEY){ // key press detected
    Serial.println(key); // print key value
    vwc.sendByte(byte(key)); // transmit key value
    delay(200); // short delay
  }
}
```

Note: there is no receiver code for this sketch because although values are being transmitted, we wish to check key presses yield the expected values. These are displayed on the serial monitor and we shall develop a simple receiver program in the next stage.

Stage 2 – Sending Key Press Values and a Simple Receiver Program

Sketch Details

We have already included the necessary code in the Tx sketch to send the values via the 433 MHz module, so for this stage we only need write a simple Rx Sketch that will receive and display the key values on the serial monitor. Note that we are receiving bytes of data that represent the ASCII codes of the pressed keys. For printing we can use char() e.g. Serial.println(char(b));

Sketch Code

The Receiver Code is as follows:

```
/*
  Expt8_Stage2_Rx - Receive data values
  Extension to stage 1 (add receiver)
*/

#include <VirtualWire.h>
#include <VWComm.h>
const int receive_pin = 2; // Rx data to D2
byte b; // byte to hold received data
String dt; // string to hold received data type
VWComm vwc; // create an object to use with VWComm

void setup()
{
  vw_set_rx_pin(receive_pin);
  vw_setup(2000);    // Bits per sec
  vw_rx_start();     // Start the receiver
  Serial.begin(9600); // initialise serial output
}

void loop()
{
  dt = vwc.dataType(); // wait for and identify data
  if (dt == "B") {
    b = vwc.readByte(); // read byte of data
    Serial.print("Byte received: "); // output text
    Serial.print(b); // read byte of data
    Serial.print(" key: "); // output text
    Serial.println(char(b)); // output character
  }
}
```

Stage 3 – A Simulated Remote Digital Lock using an RGB LED

Now, for the receiver, we introduce an RGB LED to light up in different colours to indicate different states as follows:

- Blue (steady): waiting
- Blue (flashing): key value being processed
- Green: Correct code has been entered
- Red: Incorrect code has been entered

The RGB anode pins should be connected to digital pins 2,3,4 (via suitable resistors unless you are using an RGB LED module).

Sketch Details

The secret keycode – 4321 – is stored as separate numbers in an array named keys. Remember that array elements are numbered from 0 upwards, so 4 will be stored in keys[0], 3 in keys[1] and so on. Essentially, the program gets 4 successive key values that it receives via the 433 MHz Rx and compares each one to the corresponding value in the secret code. As it does so, it sets an error count and if this is not zero after the fourth value, then the user has entered an incorrect code.

The code is simplified by using functions to perform particular tasks e.g. to set the LED colour. The function checkDigit(n) is used to receive the nth digit and compare it to the corresponding digit in the secret code. It returns a value of 0 if the digit is correct, or 1 if it is incorrect.

When all 4 digits have been received and checked, the LED is set to display GREEN if the code is correct or RED otherwise. The state is held for 10 seconds and then the program reverts to a 'waiting' state with the LED displaying BLUE.

Sketch Code

The Receiver Code is as shown overleaf:

```cpp
/*
  Expt8_Stage3_Rx - Decode signals
  Extension to stage 2 - simulate doorlock
*/

#include <VWComm.h>
#include <VirtualWire.h>
const int redPin = 2; // red led anode to D2
const int greenPin = 3; // green led anode to D3
const int bluePin = 4; // blue led anode to D4
const int receive_pin = 12; // Rx data to D12
byte keys[4] = {4, 3, 2, 1}; // Stored key code
String dt; // String to hold data type
VWComm vwc; // create an object to use with VWComm

void setup() {
  vw_set_rx_pin(receive_pin);
  vw_setup(2000); // Bits per sec
  vw_rx_start(); // Start the receiver
  pinMode(redPin, OUTPUT); // define pin as output
  pinMode(greenPin, OUTPUT); // define pin as output
  pinMode(bluePin, OUTPUT); // define pin as output
}

void loop() {
  setRGB("blue"); // indicate waiting
  byte errCount = 0; // initialise error count
  for (int i = 0; i < 4; i++) {
    // get 4 key presses
    if (checkDigit(i) > 0) {
      errCount = errCount + 1; // mismatch
    }
  }
  if (errCount < 1) {
    setRGB("green"); // indicate code correct
  }
  else {
    setRGB("red"); // indicate code not correct
  }
  delay(10000); // wait 10 seconds then start again
}

byte checkDigit(byte n) {
  //wait for value and check against stored code
  byte b; // byte to hold received key press
  byte retVal = 0; // return value = 0 if match
  dt = vwc.dataType(); // wait for and identify data
  if (dt == "B") {
```

```
    b = vwc.readByte() - 48; // convert to numeric
    if (b != keys[n]) {
      retVal = 1; // indicate mismatch
    }
  }
  flashLED("blue"); // indicate (right or wrong)
  return retVal;
}
void flashLED(String colour) {// flash led 3 times
  for (int j = 0; j < 3; j++) {
    setRGB("off");
    delay(150);
    setRGB(colour);
    delay(150);
  }
}

void setRGB(String colour) {
  // turn all pins off
  digitalWrite(redPin, LOW);
  digitalWrite(greenPin, LOW);
  digitalWrite(bluePin, LOW);
  // set selected colour
  if (colour == "red") {
    digitalWrite(redPin, HIGH);
  }
  else if (colour == "green") {
    digitalWrite(greenPin, HIGH);
  }
  else if (colour == "blue") {
    digitalWrite(bluePin, HIGH);
  }
}
```

Stage 4 – A Practical Remote Digital Lock

A low-cost electronic lock uses a simple push-pull solenoid with a 10 mm stroke. These can be obtained from a variety of sources for around £4. A typical example is as shown:

The 'bolt' (solenoid shaft) is normally in the 'out' position which would represent a locked situation when the bolt is engaged in some holding mechanism. When power is applied, the bolt retracts until power is removed. This means that the door (or other defence) needs to be opened while the bolt is retracted (which will only be for a short time). For our purposes, we will demonstrate a typical example where we hold the bolt back for 2 seconds.

The solenoid requires a 12V DC supply that can supply a continuous current of 1A, so you will need a suitable power supply. These can be obtained for around £6. You will also need a relay module to receive a command from your Arduino and switch the solenoid on or off (i.e. power on or power off). It is probably simplest to choose a model that is powered by 5V (from your Arduino) and these can be obtained for under £2 – even if this means buying a dual module!

Although the Arduino pins can only supply a maximum of around 40 mA, the load from its internal power supply is higher (around 400 mA). The relay coils use about 40 mA (but not powered from the digital pins) and have been used in many experiments without problem. If in doubt, use an external power supply – you will probably be using one anyway if this is a practical application.

Note that because this requires power to activate the unlocking, it is secure against a power disruption.

Relay modules have 3 connections: SIGNAL, Vcc, GND. The SIGNAL will be connected to a digital output pin and the others to +5V and GND respectively on the Arduino. To activate the relay, we set the digital output pin to HIGH; its normal state is LOW. The relay contacts switch between 'normally open' (N/O) and 'normally closed' (N/C) and these are indicated on the double relay module as shown:

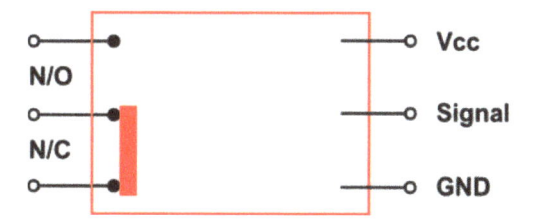

Sketch Details

In the previous stage, we caused an LED to display GREEN when a correct security code had been received and this was followed by a short delay until the system reverted to 'listening'. It is therefore a simple matter to incorporate an UNLOCK state at the same time. We simply send a signal to the relay that makes a contact that powers the solenoid. This is held for the duration of that short delay and then we signal the relay to disconnect and the solenoid returns to its locked position. For a practical application, the time delay for the unlocked position will vary according to circumstances.

Sketch Code

The previous code needs very little modification. Before the 'setup' section we need to define an output pin that will control the relay – we shall use pin 10 in this example, so we need the statement:

```
const int relayPin = 10;
```

then within the 'setup' section we want to declare that this pin is an output and initialise it to LOW:

```
pinMode(relayPin,OUTPUT);
digitalWrite(relayPin, LOW);
```

and then change the existing program to read:

```
else if (colour == "green") {
        digitalWrite(greenPin, HIGH);
        digitalWrite(relayPin, HIGH);
}
```

and finally return the relay to its 'normally off' state with:

```
digitalWrite(relayPin, LOW);
```

after the delay and before the end of the loop.

What you have Achieved and the Next Steps

In this experiment we have developed a remote locking/unlocking system that depends upon a correct 4 digit code being entered by the transmitting station. As an introduction, the sketches for transmitter and receiver have been presented as straightforwardly as possible in order that the basic principles can be readily understood and implemented. However, these might need some enhancements for particular applications and we leave this up to the reader to develop individual requirements.

It might be useful to include an LED on the transmitter that flashes when a digit has been sent and perhaps to indicate an invalid key press (non-numeric key). In turn, we might make the receiver ignore invalid characters in the key count and we leave such thoughts for the reader to explore.

Experiment 9: Using Multiple Receivers

In this experiment, we consider the situation where we have a number of devices that we wish to control remotely in one location. We can achieve this using a single transmitter and multiple receivers, each with its own ID.

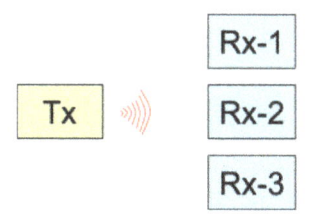

We start with a simple set-up involving a transmitter and two receivers, each of which has an LED to indicate an ON/OFF state. Further stages of the experiment introduce an LCD display and multiple buttons at the transmitter and some practical switching on the receivers.

Consideration of Technical Issues

To remotely control a number of devices in one location such as a house, we need to do a bit more than simply creating several pairs of Tx/Rx stations. This is because all transmitter and receivers are operating on the same frequency and so each receiver will try and respond to any signal that is sent from any of the transmitters. Also, there is the possibility that two transmitters could send signals at the same time and interfere with each other. Apart from the duplication of transmitters and the likely confusion of intended receiver, it is probably less convenient for the user to use multiple devices. We therefore design a system in which there is a single transmitter and multiple receivers, each of which has a designated purpose.

Probably the simplest design is for the transmitter to send a code to identify the receiver that it wishes to control and then send further data which will apply only to that station. This raises another possible problem in that this further data is also available to the other stations, so we need to ensure that they cannot misinterpret this data as signals for them. If each receiver is assigned a unique code, it would wait until it receives that code and then proceed to read the relevant number of data items that follow. All other receivers would ignore data until they received their unique identifier. One problem here is that if the code is numeric e.g. 123, then there is the possibility that this could be a data value being received by another station.

Fortunately, using the VWComm library, we are able send and receive different data types, each being identified by a code letter such as B for byte or S for String. We can therefore achieve a simple solution by identifying each receiver by a string such as Rx-1, Rx-2, etc. which acts as the key to receive further data items and act on them accordingly. Even if another receiver detects an S, it will read the subsequent information as a string that will not correspond to its unique receiver identifier.

The Experiments

Stage 1 – Simple Tx and Two Rxs

In this experiment, we demonstrate the basic principles of using multiple receivers with a transmitter that has two buttons corresponding to the two receivers identified by Rx-1 and Rx-2. The transmitter will use simple push-buttons to toggle the LEDs on the receivers.

The transmitter and receivers are connected as shown:

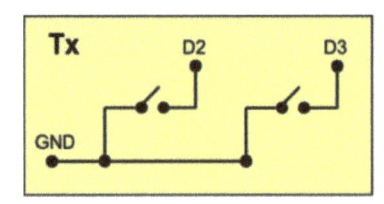

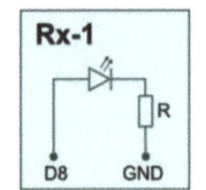

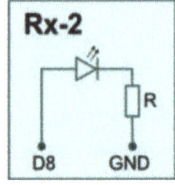

Sketch Details

The push buttons on the Tx are used to momentarily hold a digital pin LOW. In the unconnected state these will be held HIGH using the internal pull-up facility (we write a HIGH signal to the pin which has been declared as an input). The Tx uses a technique called polling which repeatedly checks the state of the input pins and acts accordingly when one goes LOW.

Each receiver waits for its identifier (Rx-1 or Rx-2) and once this is received, the LED state is toggled. The HIGH/LOW signals are equivalent to 1 and 0 respectively, so if the state is held in a boolean variable, we can toggle it with a statement like b = !b; meaning 'set b equal to NOT(b).

Sketch Code

The Transmitter Code is as follows:

```
/*
  Expt9_Stage1_Tx - Multiple receivers
  Simple test transmitter program
*/

#include <VirtualWire.h>
#include <VWComm.h>
const int transmit_pin=12; // Tx data to D12
const byte select_rx1_pin = 2; // push-button to D2
const byte select_rx2_pin = 3; // push-button to D3
VWComm vwc; //initialize an instance of the class

void setup()
{
  vw_set_tx_pin(transmit_pin);
  vw_setup(2000); // Bits per sec
  pinMode(select_rx1_pin, INPUT); // push-button input
  pinMode(select_rx2_pin, INPUT); // push-button input
  digitalWrite(select_rx1_pin, HIGH);// use pull-up
  digitalWrite(select_rx2_pin, HIGH);// use pull-up
}
void loop()
{
  if (digitalRead(select_rx1_pin) == LOW) { // Rx-1
    vwc.sendStr("Rx-1"); // send data
    delay(1000);
  }
  else if (digitalRead(select_rx2_pin) == LOW) // Rx-2
  {
    vwc.sendStr("Rx-2"); // send data
    delay(1000);
  }
}
```

The <u>Receiver Code</u> is as follows:

```
/*
  Expt9_Stage1_Rx - Multiple receivers - Rx1
  Wait for incoming text and check against ID "Rx-1"
  If it matches, toggle the state of an LED
*/

#include <VirtualWire.h>
#include <VWComm.h>
const int receive_pin = 11; // Rx data to D11
const byte led_pin = 8; // led anode to D8
String text; // string to hold received text
String dt; // string to hold received data type
boolean led_state = false; // begin with led off

VWComm vwc; // create an object to use with VWComm

void setup()
{
  vw_set_rx_pin(receive_pin);
  vw_setup(2000);   // Bits per sec
  vw_rx_start();    // Start the receiver
  pinMode(led_pin, OUTPUT); // set output pin
  digitalWrite(led_pin, led_state); // initialise led
}

void loop()
{
  dt = vwc.dataType(); // wait for incoming data
  if (dt == "S") {
    text = vwc.readStr();
    if (text == "Rx-1") { // act on Rx-1 signal
      led_state = !led_state; // toggle led on Rx-1
      digitalWrite(led_pin, led_state);
    }
  }
}
```

Stage 2 – Adding an LCD Display to the Tx

When a Tx station can perform different actions, it is useful to provide the user with some confirmation of their selection. In this stage, we use an LCD shield to provide a cheap and simple example of some basic feedback to the user.

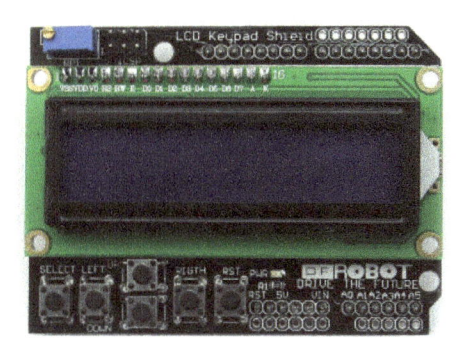

The shield plugs into the Arduino Uno and features a 16 x 2 display. It also features some push-buttons that we will use in the next stage. It can be obtained for as little as £3, but you might need to solder some headers onto it to make connections to the Arduino pins (not all of which are available). Note that on the shield we used (shown above), the available digital pins were 0 to 6 even though they appeared to correspond to 1 to 7!

Sketch Details

The previous sketch is adjusted slightly to include the Liquid Crystal library and some additional adjustments, although the essential actions are still the same. The Virtual Wire library has some default pin definitions that can conflict with the shield, so it is necessary to redefine these within the sketch. So, even though they are not used, we need to set available pin numbers for the receive pin and the press-to-transmit (ptt) pin.

The Liquid Crystal library uses a few program commands that are illustrated in the following program and most are self-explanatory. We create an object called lcd at the start of the sketch, which also defines the pins used by the display. Note the use of the lcd.begin command along with lcd.clear, lcd.setCursor, and lcd.print. When setting the cursor, the order is column, row, numbered from 0, e.g. lcd.setCursor(0,1) will cause the next print to start in the first position of the second row.

Using the LCD shield is convenient, but it does restrict the available digital pins. However, we can use the analogue pins for digital purposes and we have A1 to A5 available (A0 is used by the shield – see next stage).

Sketch Code

The Transmitter Code is as follows:

```
/*
  Expt9_Stage2_Tx - Multiple Receivers
  Adding an LCD display to the Tx
*/

#include <VirtualWire.h>
#include <VWComm.h>
#include <LiquidCrystal.h>
const byte transmit_pin = 3; // Tx data to D3
const byte receive_pin = 2; // reset default
const byte ptt_pin = 1; // reset default
const byte select_rx1_pin = A1; // push-button for Rx-1
const byte select_rx2_pin = A2; // push-button for Rx-2
VWComm vwc; // create an object to use with VWComm
LiquidCrystal lcd(8, 9, 4, 5, 6, 7);   // initialise lcd

void setup()
{
  vw_set_tx_pin(transmit_pin); // reset default
  vw_set_rx_pin(receive_pin); // reset default
  vw_set_ptt_pin(ptt_pin); // reset default
  vw_setup(2000); // Bits per sec
  pinMode(select_rx1_pin, INPUT); // set Rx-1 input
  pinMode(select_rx2_pin, INPUT); // set Rx-2 input
  digitalWrite(select_rx1_pin, HIGH);// use pull-up
  digitalWrite(select_rx2_pin, HIGH);// use pull-up
  lcd.begin(16, 2); // define lcd cols and rows
  lcd.clear(); // clear lcd display
  lcd.setCursor(0, 0); // cursor at start of line 1
  lcd.print("Waiting Input"); // output message to lcd
}

void loop()
{
  if (digitalRead(select_rx1_pin) == LOW) { // Rx-1
    vwc.sendStr("Rx-1"); // send text
    lcd.clear(); // clear lcd display
    lcd.setCursor(0, 0); // cursor at start of line 1
    lcd.print("Rx-1 Selected"); // output message
    delay(1000); // delay 1 second
```

```
  }
  else if (digitalRead(select_rx2_pin) == LOW) { // Rx-2
    vwc.sendStr("Rx-2"); // send text
    lcd.clear(); // clear lcd display
    lcd.setCursor(0, 1); // cursor to start of line 2
    lcd.print("Rx-2 Selected"); // display message
    delay(1000); // delay for 1 second
  }
}
```

Stage 3 – Adding a Keypad to the Tx

Here, we are going to simulate a keypad by using the five available buttons on the shield (the sixth is a reset button). The shield sends different values to the analogue pin A0 according to which button is pressed, thereby enabling a single pin to represent a number of different options. The sketch can read the value on A0 and interpret it accordingly using a number of pre-defined bands. Recall that analogue pins can hold values between 0 and 1023 (the analogue to digital converter uses 10 bits). For our shield, the normal state (i.e. no button pushed) is 1023. Our readings for different button pushes were 638, 406, 97, 254, and 0 for the 'select', 'left', 'up', 'down', and 'right' buttons respectively. You can check these for your shield by replacing the 'loop' code in Stage 2 with:

```
lcd.clear(); // clear lcd display
lcd.setCursor(0, 0); // cursor to start of line 1
lcd.print(analogRead(A0)); // display input value
delay(1000); // delay 1 second
```

Sketch Details

The actual values that we read on pin A0 might vary slightly, so we should allow for this by testing within a small range e.g. if(v<650 && v>630){ .... } to interpret which button is being pressed.

Note that the double ampersand (&&) indicates the logical connective AND meaning that both parts must be TRUE to produce a TRUE result.

The sketch will repeatedly read the value on A0 within the loop (a technique known as *polling* and respond to a change in the value. To detect this change, we need to store an initial value and then compare newly read values against this. To be on the safe side, we should allow for a small amount of random variation and so avoid erroneous actions.

For simplicity, we shall use just two of the buttons to select Rx-1 and Rx-2, these will be the first two buttons on the left side labelled 'select' and 'left'. This means that we are looking for values of 638 or 406 on pin A0 and let's set the boundaries as: 625-645 and 400-420 to select Rx-1 and Rx-2 respectively.

As an enhancement, we shall modify the display on the LCD display so that it shows the state of both Rx-1 and Rx-2 at any time. Recall that each time we send a signal to a given receiver, we toggle its state between ON and OFF. Both Tx and Rx start in the OFF state.

Sketch Code

These changes only affect the transmitter and the modified code is as follows:

```
/*
  Expt9_Stage3_Tx - Multiple Receivers
  Input from a digital keypad
*/
#include <VirtualWire.h>
#include <VWComm.h>
#include <LiquidCrystal.h>
const byte transmit_pin = 3; // Tx data to D3
const byte receive_pin = 2; // reset default
const byte ptt_pin = 1; // reset default
const byte button_pin = A0; // input from shield buttons
int old_btn_value = 1023; // default 'no-press'
int new_btn_value; // storage for new button value
boolean rx1_state = false; // initially Rx-1 is OFF
boolean rx2_state = false; // initially Rx-2 is OFF
VWComm vwc; //initialize an instance of the class
LiquidCrystal lcd(8, 9, 4, 5, 6, 7); // initialise lcd

void setup()
{
  vw_set_tx_pin(transmit_pin); // reset default
```

```
    vw_set_rx_pin(receive_pin); // reset default
    vw_set_ptt_pin(ptt_pin); // reset default
    vw_setup(2000); // Bits per sec
    lcd.begin(16, 2); // define lcd cols and rows
    lcd.clear(); // clear lcd display
    lcd.setCursor(0, 0); // cursor at start of line 1
    lcd.print("Rx-1: OFF"); // output message
    lcd.setCursor(0, 1); // cursor at start of line 2
    lcd.print("Rx-2: OFF"); // output message
}

void loop()
{
   // check for new button press
   new_btn_value = analogRead(button_pin);
   if (abs(new_btn_value - old_btn_value) > 10) {
     if ((new_btn_value > 625) && (new_btn_value < 645))
        {  // Rx-1 selected
           vwc.sendStr("Rx-1");
           rx1_state = !rx1_state; // toggle Rx-1 status
           lcd.setCursor(0, 0);
           if (rx1_state) {
                lcd.print("Rx-1: ON ");
           }
           else {
                lcd.print("Rx-1: OFF");
           }
        }
     else if ((new_btn_value > 400) &&
             (new_btn_value < 420))
        {  // Rx-2 selected
           vwc.sendStr("Rx-2");
           rx2_state = !rx2_state; // toggle Rx-2 status
           lcd.setCursor(0, 1);
           if (rx2_state) {
                lcd.print("Rx-2: ON ");
           }
           else {
                lcd.print("Rx-2: OFF");
           }
     }
     old_btn_value = new_btn_value;
   }
}
```

Stage 4 – Adding Relays to the Rx stations

In this last stage, we introduce some practical switching for the various receivers. While the Arduino can light up LEDs, it cannot directly switch on and off devices that use higher voltages and/or currents. However, this is easily achieved by the use of a relay in which a small current is used to control a physical switch. The relay modules are very cheap and often come in pairs. An example is shown below:

In experiment 8, we described the use of relays and so we can easily use these in the final part of this experiment. Each receiver, will be connected as shown:

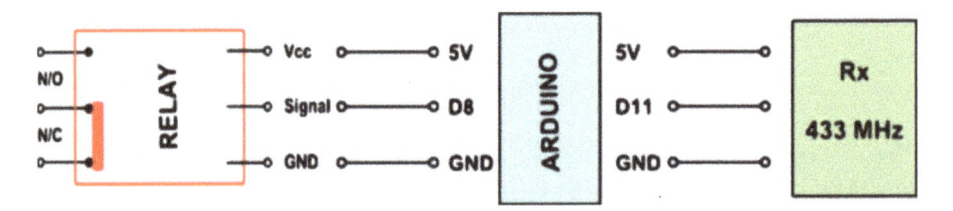

You don't need to alter the existing receiver code, although you might want to change the names of the variables *led_pin* and *led_state* to *relay_pin* and *relay_state* respectively. You can connect a separate circuit to be switched ON and OFF to the normally open (N/O) relay connections; a simple example would be a low-voltage light bulb and battery.

What You Have Achieved and the Next Steps

In this experiment we have used a single transmitter to control two or more receiving stations. Although our experiments toggled an ON/OFF state in each receiver, it would be simple to extend this to more sophisticated command signals if required. Once the receiver ID has been sent and recognised, further data could be sent without other receivers acting upon it. There is no theoretical restriction on the number of receivers although there may be practical limitations on the transmitter. Our LCD shield provides 5 available buttons and we could display the actions with careful use of the *lcd.setcursor()* command. To control more devices, it would probably be worth building a separate device with individual LEDs and an LCD display that confirms the current interaction.

There is room to improve the messaging on the LCD display as well as to use alternatives such as LEDs to indicate the selected action e.g. toggle – perhaps use RGB LEDs for red/green outputs. The use of RGB LEDs can simplify the user interface even if they are only used to display simple RED/GREEN lights corresponding to ON/OFF states.

You could extend these experiments by trying to control devices that require more than simple ON/OFF commands. For example, devise a simulator for controlling domestic appliances (although don't implement these without the guidance of qualified experts!)

Experiment 10: Remote Data Logging and Real Time Clock

In this experiment, we look at how we can receive data from a remote sensor and store the details over a period of time. Usually, we shall want to store the date and time at which these values were obtained so that we can analyse the data at a later stage. For this purpose, we require an SD card module that will communicate with our Arduino and a real-time clock (RTC) module to record the date and time.

These modules are available separately but, for convenience, we shall use a shield that combines the SD card reader and RTC functions and an 2GB SD card (each available for around £3).

The process of reading values from a sensor and storing them is known as ***data logging*** and, very often, it is more practical to transmit the sensor values to a receiver situated elsewhere. The data logging in itself is only part of a process. The data is being collected for a purpose and this means that it will eventually be analysed and/or displayed. The last part of this experiment looks at how the stored data can be accessed and used.

Consideration of Technical Issues

A typical data logging shield plugs into the Arduino and has duplicate headers to enable the user to connect to the usual digital, analogue and power pins.

Programs for the shield use the standard library modules RTClib and SD. Certain pins won't be available to the user because the shield is using these connections. However, there are still adequate free pins for most requirements and as we are typically only attaching a 433 MHz receiver to the shield, we should have no problems. Stages 1 and 2 of the experiment will provide details of pin usage.

The Real Time Clock (RTC) part of the shield requires a coin-type battery which is usually supplied with the shield. Typically, these are 3V CR1220 Lithium cells that will last years and maintain the date and clock settings even if the Arduino is disconnected.

The SD Card module will generally accept SD cards up to 2GB in capacity, which should be enough for most data logging applications. Most SD cards come pre-formatted, so they should work straight away. Generally, the cards should be formatted to FAT16 or FAT32 standards.

It should be noted that SD card modules usually operate at 3.3V, so if you are not using a shield you should power the module accordingly.

The Experiments

Stage 1 – Using the RTC Facility

This is a very simple project to demonstrate that we can get date and time values from the RTC and display them on the serial monitor.

Sketch Details

Communication between the RTC and Arduino uses a facility known as the I2C bus (although if you are using a separate module this might be different). This means that you will need to include the Wire library at the start of your program along with the RTClib. The standard connections used by I2C are the analogue pins A4 and A5, so you won't be able to use these for other purposes. As we explained in Experiment 7, the RTC has 4 connections: Vcc (to 5V), GND (to 0V), SDA (to A4) and SCL (to A5)

The sketch creates an object called RTC from which we can extract various properties such as the current day, month and year. In this sketch, a variable called *now* holds the values from reading the RTC and we are then able to access these values using now.year(), now.hour and so on.

Sketch Code

```
/*
  Expt10_Stage1_Tx - Testing the Real Time Clock
  Similar code to experiment 7
  Connect SDA to A4 and SCL to A5
*/

#include <Wire.h>
#include "RTClib.h"
RTC_DS1307 RTC; // create an RTC object
void setup () {
  Serial.begin(9600); // initialise serial output
  Wire.begin(); // create a Wire object
  RTC.begin();
  if (! RTC.isrunning()) {
    Serial.println("RTC is NOT running!");
    RTC.adjust(DateTime(__DATE__, __TIME__));
  }
}

void loop () {
  DateTime now = RTC.now(); // get date and time
  Serial.print(now.year(), DEC); // output date values
  Serial.print('/');
  Serial.print(now.month(), DEC);
  Serial.print('/');
  Serial.print(now.day(), DEC);
  Serial.print(' ');
  Serial.print(now.hour(), DEC); // output time values
  Serial.print(':');
  Serial.print(now.minute(), DEC);
  Serial.print(':');
  Serial.println(now.second(), DEC);
  delay(3000); // delay 3 seconds
}
```

Stage 2 – Writing and Reading to/from an SD Card

Sketch Details

It is suggested that, before you start this stage, you load the *CardInfo* sketch from the SD section of *examples* in the Arduino IDE. If you look at the first part of the program, you will see a commented section that refers to the CS pin. This connection varies with different shields, so you may need to experiment by changing this value (4, 8 or 10 are typical). Once this is successfully established, you will be able to view the details of your SD card on the serial monitor and you can then put the correct CS pin details in the following program.

Communication between the SD module and Arduino uses the SPI bus which has standard connections labelled MISO, MOSI, SCK, and CS connected to digital pins 12, 11, 13, and 10 respectively (although CS may vary – see above). Again, this means that you should not use these pins for other purposes.

In our test program, we shall start by writing a sequence containing text and values and then read them from the SD card while displaying the values on the serial monitor. Note that if the file already exists, the data will be appended to that already in the file. This is a typical requirement for data logging if the interval between reading data values is not short and there is merit in closing the file between readings so as retain file integrity in the event of a problem. For the rapid collection of data, the file can remain open for the duration of the exercise although care should be taken to not exceed the capacity of the SD card. If a new data set is required, the file should be deleted prior to starting the experiment – this is most easily done on a PC.

Note that if you do want to replace an existing file when writing data, you can use the command SD.remove("filename") to delete a file following an if(SD.exists "filename").

You may notice that two sets of values are written to your file when you run the program. This is because the program runs as soon as it is loaded but is forced to restart when you open the serial monitor.

Sketch Code

```
/*
  Expt10_Stage2_Rx - Testing data written an SD Card
*/

#include <SPI.h>
#include <SD.h>
const int CSPin = 10; // data pin of SD card to D10
File testFile; // create a File object
String filename = "test.txt"; // name of text file

void setup()
{
  Serial.begin(9600); // initialise serial output
  if (!SD.begin(CSPin)) { // initialise SD Card
    Serial.println("Could not initialise SD Card");
    return;
  }
  // open file for writing data
  testFile = SD.open(filename, FILE_WRITE); // open file
  if (testFile) {
    Serial.println("Writing data"); // message to user
    testFile.println("testing 1, 2, 3."); // Sample data
    testFile.close(); // close file
  }
  else {
    // if the file didn't open, print an error:
    Serial.print("error opening file: ");
    Serial.println(filename);
  }
  // re-open the file for reading data
  testFile = SD.open(filename, FILE_READ);
  if (testFile) {
    while (testFile.available()) {
      Serial.write(testFile.read()); // output file data
    }
    testFile.close();
  } else {
    Serial.println("Could not open file for reading");
  }
}

void loop()
{
        // no code here
}
```

Stage 3 – Transmitting, Receiving and Storing Data and Time Values

Sketch Notes

This experiment combines a sensor with a Tx module and sends data at regular intervals. The RTC module is located at the receiver end and is used to store date and time values along with the received data. The data file is opened and closed for each reading and successive values are appended to the file. It is not advisable to stop the receiver program while the file is open and so we need a plan for terminating the process without corrupting the file. We present the scenario where the transmitter will continue to send values ad infinitum, but that at some stage we wish to take the values stored on the SD card and analyse them (also allowing for the process to resume at a later stage with a new data file).

Various solutions can be devised according to the particular project details. In this experiment, we are sending data every 10 seconds, so one solution is to give a visual indication of when it is safe to disconnect the receiver. Our RGB LED module can be used here to display:

- 🔴 RED – do not disconnect
- 🟢 GREEN – OK to disconnect
- 🔵 BLUE – about to open file.

So while the LED is displaying GREEN, the receiver can be disconnected and there is a safety zone while the LED is showing BLUE just in case you disconnect at the very end of the GREEN time.

The sensor that we shall use here is the DHT 11 temperature/humidity sensor that we used in our weather station in Experiment 7. A pull-up resistor (4.7KΩ - 10KΩ) is recommended.

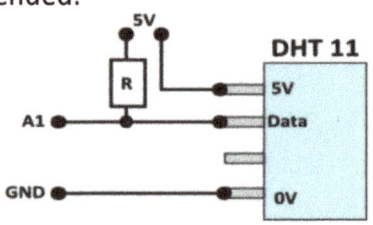

Sketch Code

The Transmitter Code is as follows:

```
/*
  Expt10_Stage3_Tx - Sending Data Values
  Humidity and Temperature from DHT11
*/

#include <VWComm.h>
#include <dht.h>
#include <VirtualWire.h>
const int transmit_pin = 12; // Tx data to D12
const int transmit_en_pin = 3; // reset default pin
const int dht_dpin = A1; // DHT11 data to A1
int x; // variable for data values
dht DHT; // create a dht object
VWComm vwc; // create an object to use with VWComm

void setup()
{
  vw_set_tx_pin(transmit_pin); // Tx data pin
  vw_set_ptt_pin(transmit_en_pin); // reset default
  vw_setup(2000); // Bits per sec
  Serial.begin(9600); int x; // initialise serial output
  Serial.println("Starting …"); // indicate start
}

void loop()
{
  vwc.sendStr("Synch"); // synchronise with receiver
  DHT.read11(dht_dpin); // get data values from DHT11
  x = int(DHT.humidity); // read humidity value
  vwc.sendStr("Humidity:"); // send text
  Serial.print("Humidity: "); // echo text to monitor
  vwc.sendInt(x); // send data value
  Serial.println(x); // echo data value to monitor
  x = int(DHT.temperature); // read temperature value
  vwc.sendStr("Temperature:"); // send text
  Serial.print("Temperature: "); // echo text to monitor
  vwc.sendInt(x); // send data value
  Serial.println(x); // echo data value to monitor
  delay(10000);  // Pause for 10 seconds.
}
```

The Receiver Code is shown overleaf:

```
/*
  Expt10_Stage3_Rx - Receiving Data Values
  and storing them on an SD card with date/time
*/

#include <Wire.h>
#include <RTClib.h>
#include <VirtualWire.h>
#include <VWComm.h>
#include <SPI.h>
#include <SD.h>

const int receive_pin = 7; // Rx data to D7
const int CSPin = 10; // Set Chip Select for this shield
const int red = 2; // D2 connected to red of RGB LED
const int green = 3; // D3 connected to green of RGB LED
const int blue = 4; // D4 connected to blue of RGB LED
int i; // variable to store received integer values
String dt; // string to identify received data type
String text; // string to hold received text
String filename = "test.txt"; // filename for data
File testFile; // create a File object
VWComm vwc; // create an object to use with VWComm
RTC_DS1307 RTC; // create an RTC object

void setup()
{
  vw_set_rx_pin(receive_pin); // set Rx data pin
  vw_setup(2000); // Bits per sec
  vw_rx_start(); // initialise receiver
  Serial.begin(9600); // initialise serial output
  Wire.begin(); // initialise Wire
  RTC.begin(); // initialise RTC
  if (! RTC.isrunning()) { // set date/time values?
    Serial.println("RTC is NOT running!");
    RTC.adjust(DateTime(__DATE__, __TIME__));
  }
  if (!SD.begin(CSPin)) { // initialise SD card
    Serial.println("Could not initialise SD Card");
    return;
  }
  Serial.println("Starting..."); // Inform user
  setRGBLed("green"); // indicate no active transfer
}

void loop()
{
  text = ""; // initialise text string
```

```
  do {
    dt = vwc.dataType(); // wait for and identify data
    if (dt == "S") {
      text = vwc.readStr(); // get incoming text
    }
  } while (text != "Synch"); // wait for synch signal

  setRGBLed("red"); // receiving and recording data
  appendToFile(GetDateTime());
  for (int j = 0; j < 2; j++) {// get humidity and temp
    dt = vwc.dataType(); // wait for and identify data
    if (dt == "S") {
      text = vwc.readStr(); // get text
      appendToFile(text); // store text in file
    }
    else
    {
      appendToFile("???"); // unexpected data item
    }
    dt = vwc.dataType(); // wait for and identify data
    if (dt == "I") {
      i = vwc.readInt(); // get data
      appendToFile(String(i)); // store data in file
    }
    else
    {
      appendToFile("???"); // unexpected data item
    }
  }
  setRGBLed("green"); // indicate waiting for new data
}
String GetDateTime() { // create date/time string
  String dtval = "";
  DateTime now = RTC.now();
  dtval = String(now.day()) + "/";
  dtval = dtval + String(now.month()) + "/";
  dtval = dtval + String(now.year()) + " ";
  dtval = dtval + String(now.hour()) + ":";
  dtval = dtval + String(now.minute()) + ":";
  dtval = dtval + String(now.second());
  return dtval;
}
void appendToFile(String text) // add data/text to file
{
  testFile = SD.open(filename, FILE_WRITE);
  if (testFile) {
    testFile.println(text);
```

```
    testFile.close();
  }
}
void setRGBLed(String colour) { // display chosen colour
  // set all RGB LED pins LOW
  digitalWrite(red, LOW);
  digitalWrite(green, LOW);
  digitalWrite(blue, LOW);
  //set the selected colour(s) HIGH
  if (colour == "red") {
    digitalWrite(red, HIGH);
  }
  else if (colour == "green") {
    digitalWrite(green, HIGH);
  }
  else if (colour == "blue") {
    digitalWrite(blue, HIGH);
  }
  // any other colour value results in OFF
}
```

Stage 4 – Processing the Stored Data

You now have data stored in a text file on your SD card and you can use any text editor to view the contents of your file – you should be able to just double-click on the file icon to view the details. A typical sample (from Stage 3) is shown:

> 11/3/2016 17:19:38
> Humidity:
> 38
> Temperature:
> 19
> 11/3/2016 17:19:50
> Humidity:
> 39
> Temperature:
> 19

Note that the text for 'humidity' and 'temperature' is not really necessary if we know the sequence of the data, so we could modify the transmitter sketch to not send these. Also, in the receiver sketch, we might want to store the date and time as separate items if it simplifies the processing of data at a later stage.

The actual processing of the data will depend upon your preferences and expertise. Microsoft Excel is a typical candidate and it is worth noting that you can read data directly from your Arduino, via the serial port, into Excel using a plug-in called KeyStroke Reader. However, you can import the data from your SD card using the "data from text file" option in Excel. For this, we need to make a few small changes to our Rx sketch so that the data appears as:

31/3/2016,18:33:47,38,19,

The data values for each reading are separated by commas (other separators can be used) so that Excel will place them in different cells. The last comma is not necessary, but it makes no difference and simplifies the program.

At the start of the program we declare a string variable ftext that will hold each set of readings:

```
String ftext;
```

And then the loop code becomes:

```
void loop()
{
  text = ""; // initialise text string
  do {
    dt = vwc.dataType(); // wait for and identify data
    if (dt == "S") {
      text = vwc.readStr(); // get incoming text
    }
  } while (text != "Synch"); // wait for synch signal
  setRGBLed("red"); // receiving and recording data
  ftext = GetDateTime() + ",";
  for (int j = 0; j < 2; j++) {// get humidity and temp
    if (dt == "I") {
      i = vwc.readInt(); // read data value
      ftext = ftext + String(i) + ",";
    }
  }
  appendToFile(ftext); // store data in file
```

```
    setRGBLed("green"); // indicate waiting for new data
}
```

Finally change the line in the function GetDateTime() to read:

```
    dtval = dtval + String(now.year()) + ",";
```

Another possibility is to use the 'Processing' program that was the forerunner to the Arduino IDE and presents a similar user interface. It is freely available at https://processing.org/download/ and there is a useful guide called 'Getting Started with Processing' by Reas and Fry that is a valuable, easy-to-read book that explains how to use its many facilities.

If you have programming skills, then you can read data from the file and do whatever you want with it, subject to the limitations of your programming language.

What you have Achieved and the Next Steps

In this experiment we have read values, at regular intervals, from a sensor and sent them to a receiver which, in turn, has stored the values on an SD card along with a time-date stamp. We have made sure that the receiver is synchronised with the transmitter by waiting for a given 'Synch' signal from the transmitter. We have considered the final aspects of processing the data and suggested some possibilities, but as these are so diverse, we leave it to the reader to employ their own particular skills for this purpose.

It should be noted that it is not always necessary to store data and then process it; values can be read and displayed in real-time over the internet and there are free ways of doing this. The reader is referred to books such as 'Exploring Arduino' by Jeremy Blum and 'Arduino Workshop' by John Boxall for further details.

Final Remarks

This set of tutorial projects has been designed as a structured approach to take you through a series of remote control experiments that gradually increase in complexity. We have concentrated on some readily available 433 MHz Tx/Rx modules that have a reasonable range for domestic use, using the aerials suggested.

The experiments and accompanying code should enable you to write your own sketches for your particular requirements. In the later experiments, we have tried to emphasise the importance of synchronising communication between transmitter and receiver, as well as writing robust code that allows for errors and deals with them in a way that keeps the user informed and allows for re-transmission.

For any project, rigorous testing needs to be undertaken to ensure that actual and expected results coincide! Always run short tests with any new sensor to ensure that the received values are consistent and accurate. Also, try to build any new sketch in a way that introduces new features, one by one, so that they can be tested and evaluated individually.

While the experiments cover a range of typical applications, we have not attempted to deal with the remote control of model cars etc. This is an area in which more complicated components are required and signalling requirements are somewhat different. Importantly, the transmission range of the 433 MHz modules used in this book may be insufficient, even with the use of aerials.

One of the projects involves a remote-controlled door lock - see experiment 8. When creating similar projects, the user should be aware that the signals can also be detected by other receivers and cloning devices. If security is likely to be an issue, then it is suggested that the programs should employ some tactics to avoid compromise e.g. by cycling through a number of different security codes.

APPENDIX A - COMPONENT LIST

Temperature and Humidity Sensor (DHT11)

80p (Banggood) - £2.48 (Amazon)

Pressure and Temperature Sensor (BMP180)

99p - £2.99 (eBay)

433 Mhz Transmitter & Receiver

£1.02 (Amazon) - £1.38 (MinilmTheBox)

Push Button

100 pcs £1.51 (Banggood) - £2.99 (Amazon)

Light Emitting Diode (single colour LED)

10 each of 5 colours £1.79 (eBay) - 8p (various)

Light Emitting Diode Module (tri-colour LED)

£1.94 (EachBuyer.com) - £2.36 (Banggood)

Resistors (10K ohms)

10pcs 15p (various)

Real Time Clock (DS1307 or similar)

£1.18 (MinilnTheBox) - £2.55 (eBay)

COMPONENT LIST - Continued

20x4 Liquid Crystal Display (LCD) £3.77 (eBay) - £17.00 (Makersify)	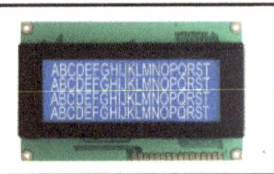
16x2 Liquid Crystal Display (LCD Shield) £2.52 - £7.99 (Amazon)	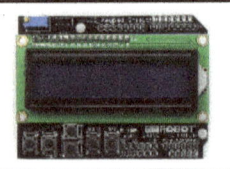
Connectors (Male to Male) Typical: 65 pcs, 11 - 26cm £2.50 (Amazon)	
Connectors (Male to Female) Typical: 40 pcs, 10cm £2.50 (Amazon)	
Spiral Spring Aerial 10 pcs £1.59 (Banggood) - £2.69 (Amazon)	
Whip Aerial £3.94 (Rapid Electronics) - £9.78 (RS Components)	
Buzzer 2pcs £1.99 (Squirrel-labs UK) - £4.13 (Amazon)	
Keypad 1pc £1.99 - 5pcs £2.99 (both through Amazon)	

COMPONENT LIST - Continued

Relay From £1.20 through Amazon suppliers	
Electronic Lock From £ 2 + dependent on use required	
Power Supply From £3.58 (Accmart)	
Arduino Uno £3.84 (GearBest) - £24.99 (Maplin Electronics)	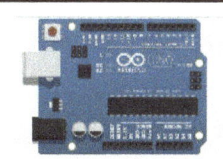
Header Connectors Many suppliers	
Loud Buzzer Module £2.08 (Amazon)	
Data Logging Shield Approx. £4 (Amazon)	
SD Card Module (option to use with RTC) Approx. £1.50 (Amazon)	

COMPONENT LIST - Continued

Important Notes

The prices given are only a guide price and higher and lower prices can frequently be found.

Many of the cheaper suppliers are located in China and therefore delivery times may be considerable (up to several weeks) and therefore a supplier based in your own country is often preferable if you want your components quickly.

The fact that a supplier is named in this list is not indicative of an endorsement of its reliability or the up to date cost of the component.

Many of the components and Arduino boards are supplied with the headers not soldered in place. This does give you the flexibility in whether male or female, straight or angled headers are fitted. Note that if a shield needs headers soldering on they must be at 90 degrees to the board and chosen to allow the shield to be mounted on the Arduino Uno.

Use a fine tipped soldering iron and a good quality multicore solder. Check for accidental solder bridges with a multimeter or continuity tester before using the component especially when soldering headers to a shield or the Arduino Uno. It is also worth checking the header connections on new components when you receive them, as sometimes the soldering is poor. It is easier to check this first before you completely rewrite a programme you have written because it is not working!

APPENDIX B - THE VWComm LIBRARY

The VWComm library is used alongside *Virtual Wire* to provide a simple set of commands for sending and receiving items of data including numbers and text. It was created by the authors of this book to simplify the coding of the various remote-control projects based upon the 433 MHz modules used in these chapters. It can be freely downloaded from this book's companion website:

www.learningbyprojects.uk

The library is available in a zip folder and can be installed in the usual way (see the section on library files in the chapter 'Writing Sketches' at the start of this book).

Usage

All sketches need to include the following code at the start:

```
#include <VirtualWire.h>
#include <VWComm.h>
VWComm vwc;
```

Also, your setup code should include the normal entries associated with the Virtual Wire library e.g.

```
  vw_set_rx_pin(receive_pin);
  vw_setup(2000); // Bits per sec
  vw_rx_start();
```

VWComm allows you to send and receive the following data types: byte, integer, float, and text. The relevant commands are given below:

The Transmitter Code can use the following commands:

```
vwc.sendByte(b); // send value of b (0-255)
vwc.sendInt(i); // send value of i (+/- 32767)
vwc.sendFloat(x, 4); // send f.p. value with 4 d.p.
vwc.sendStr(s); // send text string stored in s
```

It is suggested that, when sending data, a short delay e.g. 500 milliseconds, follows each data item in order to allow time for the receiver to process the incoming data and be prepared to receive more.

The Receiver Code can use the following command:

```
dt = vwc.dataType(); // wait for incoming data
```

In the above, dt is a string variable and the command causes the sketch to wait for a signal from the transmitter that contains a value for 'B', 'I', 'F' and 'S' to indicate the data type that follows. The receiver then checks the value of dt to call the appropriate routine to decode the particular data with commands such as:

```
if (dt == "S") {
   text = vwc.readStr();
        ... code ...
   }
```

The reader is referred to experiment 4 for examples. Also, be aware that Virtual Wire uses default pin numbers for facilities that may not be required e.g. ptt_pin, however these may conflict with pins used by standard shields and may therefore need redefining in sketch code.

APPENDIX C - GLOSSARY OF TERMS

#define: a constant value is given a name before the sketch is compiled.

#include: used to include libraries in the sketch.

Aerial: a rod, wire or structure by which signals are transmitted or received as part of a Tx/Rx setup.

Analogue Data: a continuous variable e.g. voltage or temperature.

analogRead(): reads the value of a pin and converts it to a digital value (between 0 and 1023).

analogWrite(): produces a square wave of a specified frequency.

Antenna: see aerial.

Arduino: an open-source electronic prototyping platform - see *www.arduino.cc*

Array: a data structure in which a list can be stored and referenced by an index number.

ASCII Code: a number code for characters (originally 0-127).

Baud Rate: the rate at which information is transferred (bits/sec).

Bit short for *binary digit* (0 or 1). The basic unit of information.

boolean: a data type for variables that have values TRUE or FALSE corresponding to 1 and 0 respectively.

Bootloader: software loaded onto the ATMega at manufacturing time to enable user program downloads via the USART.

Bounce: the rapid on and off that occurs as a mechanical switch is either opened or closed.

Bus: a set of connections between devices that carry data and control signals.

byte: a data type for variables storing numbers between 0 and 255.

C++: a popular programming language (used by Arduino sketches).

Class: a container for data and functions as part of an application.

COM: communication port. A type of serial port associated with a number assigned by Windows.

const: a keyword that effectively makes a variable read-only.

Control structures: statements such as "if", "else", "goto" etc used in programming languages to determine the flow of program logic.

Data logging: data is read from sensor(s) and stored for later analysis and or display.

Debugging: finding out why your sketch doesn't do what you wanted it to!

delay(m): causes a sketch to pause for m milliseconds.

Digital data: signals that vary in discrete steps as opposed to analogue signals whose values vary in a continuous manner.

digitalRead: reads a HIGH or LOW value from a digital pin

digitalWrite: writes a HIGH or LOW value to a digital pin

double: normally a double is a more precise variable than a float. However on the Arduino it is the same.

FAT16 or 32: formatting systems for memory cards and disks.

Flash memory: computer data storage system that can be erased and reprogrammed.

float: a variable used instead of an integer where a decimal point is needed.

Frequency: rate per second that a wave occurs.

Function: segment of programming code that performs a defined task.

Ground (GND): circuit return, corresponds to the negative and completes a circuit.

Handshake: a signalling process that coordinates communication between devices.

Header: male or female connectors.

IDE: integrated development environment - a piece of software used by programmers to develop applications.

Input: data from a sensor that can be processed.

int: defines an integer variable using 2 bytes (4 on the Arduino Due). A 2-byte integer can store numbers between -32768 and +32767.

Interrupt: a temporary suspension of the programme code to run a service routine.

Invert: a term used to indicate the reversal of a binary state.

LCD: liquid crystal display.

LED: light emitting diode (the longer pin is the anode (+)).

Library: external code included on compilation that simplifies a sketch e.g. handling hardware and manipulating data (see #include).

Loop: a section of program code that can be executed repeatedly.

ON/OFF: corresponds to the logic level 1 and 0, HIGH or LOW.

pinMode(): configures a specified pin to behave as an input or an output.

Polling: a monitoring technique to repeatedly check the inputs and respond accordingly.

Potentiometer: variable resistor.

Program: a set of instructions to complete a given task.

Protocol: a set of rules used when computers communicate with each other.

Pull-down resistor: connected between ground and a signal input that ensures that the default state is LOW. Typically 10 KΩ.

Pull-up resistor: connected between the positive supply and a signal input that ensures that the default state is HIGH. Typically 10 KΩ.

Pulse-width modulation (PWM): a simulated analogue output on a digital pin by producing a square wave of a specified frequency and variable pulse width.

RAM: Random Access Memory - a type of data storage in which access time is not dependent on physical location of the data.

Range: upper and lower limits of data.

Real time clock (RTC): a module that keeps time using a small on board battery.

Relay: uses a small current output to switch on or off larger current devices.

Resistor: in a circuit resistors reduce current flow and may act to reduce the voltage in a circuit.

Rx: the receiver in a transmit/receive setup.

SD card: Secure Digital (SD) cards are a standard format system of storing data.

Sensor: detects events or environmental changes and outputs a signal to the Arduino (e.g. light or temperature).

Serial: a method of communication between devices.

setup(): part of an Arduino sketch that is only executed at start up.

Shield: a board that piggy-backs the Arduino to give extra functionality (e.g. LCD shield, WiFi shield).

Solenoid: voltage applied produces a magnetic field (used in electrically operated locking devices).

String: a sequence of characters. Also a data type.

Toggle: a technique where different states are achieved by the same action e.g. press switch - ON , press switch again- OFF

tone(): a program command to generate a square wave of a specified frequency.

Tx: the transmitter part of an transmit/receive setup

UART: Universal Asynchronous Receiver/Transmitter - hardware to facilitate serial communication.

USB: Universal Serial Bus - a standard for connecting devices to a computer.

Variable: a name associated with a value stored by a program.

volatile: a compiler directive used for variables accessed from interrupt service routines.

Wavelength: length between adjacent peaks or troughs in a waveform.

Word: a term used to refer to a 'working' unit of storage, usually equal to the size of data used by the processor. For the Arduino, this is either 16 or 32 bits dependent on the board used.

APPENDIX D - STANDARD TEMPLATE FILES

Within this book we have devised experiments based upon some standard templates using the VWComm library for the Tx and Rx sketches. These basic templates are available on the companion website for this book along with the VWComm library - see:

> www.learningbyprojects.uk

While the basic templates involve output to the serial monitor, we have also developed projects that have used output to an LCD display (both stand-alone and shield versions) as well as to data files via an SC card shield or module. We have therefore created some additional template files as 'generic receiver modules' for the user to download and modify as required.

BIBLIOGRAPHY

Books:

Bohall, J. 'Arduino Workshop', No Starch Press, 2013

Blumm, J. 'Exploring Arduino', Wiley, 2013

Monk, S. 'Programming Arduino - Getting Started with Sketches', McGraw-Hill, 2012

Monk, S. 'Programming Arduino - Next Steps', McGraw-Hill, 2014

Reas, C. & Fry, B. 'Getting Started with Processing', O'Reilly, 2010

Websites (Reference):

https://www.arduino.cc/en/Reference/HomePage

http://www.eng.utah.edu/~cs5780/debouncing.pdf

http://www.pulseelectronics.com/products/antennas/antenna_basic_concepts

http://www.qrz.ru/schemes/contribute/arrl/chap18.pdf

https://arduino-info.wikispaces.com/DHT11-Humidity-TempSensor

https://learn.sparkfun.com/tutorials/bmp180-barometric-pressure-sensor-hookup-

http://www.airspayce.com/mikem/arduino/VirtualWire.pdf

https://github.com/sui77/rc-switch/wiki

http://physics.tutorvista.com/waves/wavelength.html

http://www.tpub.com/neets/book10/40j.htm

Datasheets:

http://www.micropik.com/PDF/dht11.pdf

https://cdn-shop.adafruit.com/datasheets/BST-BMP180-DS000-09.pdf

Resources:

http://www.learningbyprojects.uk

https://processing.org/download/?Processing

https://www.arduino.cc/en/Main/Software

Library	Source
VirtualWire.h	github.com/danielesteban/ArduinoLib/blob/master/VirtualWire/VirtualWire.h
VWComm.h	See companion website for this book
dht.h	http://playground.arduino.cc/Main/DHTLib
SFE_BMP180.h	github.com/sparkfun/BMP180_Breakout_Arduino_Library/archive/master.zip
Wire.h	Standard Arduino library
RTClib.h	github.com/adafruit/RTClib/blob/master/RTClib.h
LiquidCrystal.h	Standard Arduino library
Keypad.h	playground.arduino.cc/Code/Keypad
SPI.h	Standard Arduino library
SD.h	Standard Arduino library

NOTES

NOTES

NOTES